Springer Tracts in Advanced Robotics

Volume 8

Editors: Bruno Siciliano · Oussama Khatib · Frans Groen

Springer-Verlag Berlin Heidelberg GmbH

Engineering ONLINE LIBRARY

http://www.springer.de/engine/

J. Baeten · J. De Schutter

Integrated Visual Servoing and Force Control

The Task Frame Approach

With 115 Figures

Springer

Professor Bruno Siciliano, Dipartimento di Informatica e Sistemistica, Università degli Studi di Napoli Federico II, Via Claudio 21, 80125 Napoli, Italy, email: siciliano@unina.it

Professor Oussama Khatib, Robotics Laboratory, Department of Computer Science, Stanford University, Stanford, CA 94305-9010, USA, email: khatib@cs.stanford.edu

Professor Frans Groen, Department of Computer Science, Universiteit van Amsterdam, Kruislaan 403, 1098 SJ Amsterdam, The Netherlands, email: groen@science.uva.nl

STAR (Springer Tracts in Advanced Robotics) has been promoted under the auspices of EURON (European Robotics Research Network)

Authors

Prof. Johan Baeten
Katholieke Hogeschool Limburg
Department of Industrial Sciences and Technology, Division Automation
Universitaire Campus Gebouw B, bus 3
B-3590 Diepenbeek, Belgium
E-mail: johan.baeten@iwt.khlim.be
URL: www.khlim.be/~jbaeten

Prof. Joris De Schutter
Katholieke Universiteit Leuven
Department of Mechanical Engineering
Division of Production Engineering
Machine Design & Automation (PMA)
Celestijnenlaan 300B
B-3001 Leuven (Heverlee), Belgium
E-mail: joris.deschutter@mech.kuleuven.ac.be
URL: people.mech.kuleuven.ac.be/~jdeschut

ISSN 1610-7438
ISBN 978-3-642-07340-3

Library of Congress Cataloging-in-Publication Data
Baeten, J. (Johan), 1968-
Integrated visual servoing and force control : the task frame approach / J. Baeten, J. De Schutter.
p. cm. -- (Springer tracts in advanced robotics ; 8)
Includes bibliographical references and index.
ISBN 978-3-642-07340-3 ISBN 978-3-540-45057-3 (eBook)
DOI 10.1007/978-3-540-45057-3

1. Robotics--Control systems. 2. Servomechanisms. I. Schutter, J. de (Joris) II. Title. III. Series.

TJ211.35.B34 2003
629.8'92--dc22 2003059533

http://www.springer.de

Originally published by Springer-Verlag Berlin Heidelberg New York in 2004
Softcover reprint of the hardcover 1st edition 2004

Typesetting: Digital data supplied by author.
Data-conversion and production: PTP-Berlin Protago-TeX-Production GmbH, Berlin
Cover-Design: design & production GmbH, Heidelberg
Printed on acid-free paper 62/3020Yu - 5 4 3 2 1 0

Foreword

At the dawn of the new millennium, robotics is undergoing a major transformation in scope and dimension. From a largely dominant industrial focus, robotics is rapidly expanding into the challenges of unstructured environments. Interacting with, assisting, serving, and exploring with humans, the emerging robots will increasingly touch people and their lives.

The goal of the new series of *Springer Tracts in Advanced Robotics (STAR)* is to bring, in a timely fashion, the latest advances and developments in robotics on the basis of their significance and quality. It is our hope that the wider dissemination of research developments will stimulate more exchanges and collaborations among the research community and contribute to further advancement of this rapidly growing field.

The monograph written by Johan Baeten and Joris De Schutter builds upon an important paradigm for sensor-based robotics inspired by human natural behaviours: see and touch! An integrated vision/force approach is developed which clearly bridges the existing gap between visual and force feedback, up to the point of keenly combining the advantages of both sensors while overcoming their individual drawbacks. Its potential impact on the performance of the next generation of robots is starting to be recognized by major manufacturers world-wide. To this regard, the unique feature of the work lies in its comprehensive treatment of the problem from the theoretical development of the various schemes down to the real-time implementation of interaction control algorithms on an industrial robot.

Remarkably, the doctoral thesis at the basis of this monograph was a finalist for the Second EURON Georges Giralt PhD Award devoted to the best PhD thesis in Robotics in Europe. A fine addition to the series!

Naples, Italy
May 2003

Bruno Siciliano
STAR Editor

to Jos Simons
(1951 – 2002)

Preface

Sight and touch are two elementary, but highly complementary, senses. They contribute to the unique abilities of one of the most versatile manipulators of nature: man. Projecting these sensing abilities to robotics fans our imagination of the robot as an image of human kind. Whenever one of these senses is missing, we might find ourself in the dark or feel isolated. Is this what a 'sensorless' robot must feel like? Then, let him see and let him feel!

Early research on vision and force sensing created two separate interest groups: the visual servoing community and the force control community. Due to the very complementary nature of vision and force, both groups developed very specific ways in tackling sensor based robotics. Initially, this work was situated in some kind of niche, with almost no references to integrated vision/force approaches in robotics. Clearly, the lack of a common basis for integrated vision and force control prevented the bridging of both worlds. Bridging this gap is the aim of this book. The framework presented in this book provides and answer to the recently emerging demand for an integrated vision/force approach. This emerging demand is not surprising, since the development of combined vision/force applications is only a natural step in the evolution of sensor based robotics.

Evidently, the integrated vision/force approach combines the advantages of both sensors, while overcoming their individual drawbacks. Moreover, in contrast to what one might expect, the effort needed to accomplish a combined vision/force task is less than the sum of the efforts needed for both visual servoing and force control applied separately. After all, avoiding collision or depth estimation in a pure visual servoing task or detecting a change in the contact situation, e.g. at a corner, in a force controlled task are annoying operations which are hard to deal with. Using both sensors together remedies these shortcomings. In fact, this brings us back to the comparison with the human manipulator. We humans seamlessly switch or choose between touch and vision to accomplish a wide range of tasks, following one golden rule of thumb – true for both humans and sensor based robotics – : 'Assign specific tasks to that sensing ability which is best suited to perform the task'.

At this point, I'd like to express my gratitude to my supervisors and colleagues of the department of Industrial Sciences and Technology at the Katholieke Hogeschool Limburg, for the time and opportunity offered, and to my supervisors and colleagues of the department of Mechanical Engineering at the the Katholieke Universiteit Leuven, for their support and guidance. I'd further like to thank prof. Luc Vangool, prof. Hendrik Van Brussel, prof. Jan Swevers and prof. Herman Bruyninckx from the K.U.Leuven, and prof. Bruno Siciliano from the Università degli Studi di Napoli Federico II (Italy) for their efforts to my account.

Sint-Truiden, April 2003 *Johan Baeten*

This monograph is based on the PhD research work carried out by Johan Baeten under my supervision. It is the result of five years of fruitful collaboration (1996-2001) during which Johan was both a lecturer at the Katholieke Hogeschool Limburg and a part-time researcher at K.U.Leuven. In this work we tried to bring together force control and visual servoing in one common control framework and examine potential applications. We gratefully acknowledge financial support from the Belgian Interuniversity Poles of Attraction Programme, initiated by the Belgian State, Prime Minister's Office, Science Programming and from the K.U.Leuven Concerted Research Action GOA/99/04.

Johan's PhD thesis was selected as a finalist of the EURON Georges Giralt PhD Award - Second Edition. Of course we are very proud and honoured that this work was appreciated by the EURON PhD Award selection committee and we are very grateful for the offer to give it wider exposure by the publication of this book.

Research on robot force control at the Division of Production Engineering, Machine Design and Automation (PMA) of K.U.Leuven started in 1974 under the direction of Professor Hendrik Van Brussel. Jos Simons was his first PhD student. In 1980 he finished his PhD on the *Active Adaptable Compliance Wrist (AACW)*, which was proposed as a universal solution for the force feedback problem in robotic assembly. In his (3D) experiments Jos introduced adaptable passive compliance and combined it with active force feedback. The work even included learning behaviour! After obtaining his PhD, Jos spent one year at the Jet Propulsion Lab in Pasadena working on off-line robot programming. Back in Leuven in 1981, he interfaced a Cincinnati-T3 Milacron robot to a HP1000 minicomputer, wrote an off-line programming library, and started experiments with real-time force control and (Kalman filter based) visual servoing. This work was done in collaboration with Jayantha Katupitiya (PhD in 1985, now at the University of New South Wales in Sydney) for the visual servoing part, and myself for the force control part (PhD in 1986). Jos left the university in 1983 for a job in industry. Unfortunately, he passed away just a few months ago, in September 2002, at the age of 51. Twenty years later we can only conclude that Jos was a true robotics pioneer and visionary, and that by changing over to industry he missed a brilliant career as a robotics researcher. The publication of this book is the right occasion to pay tribute to him. The work presented in this book is based on his early ideas, approaches and experiments, which continue to inspire us every day. Therefore we dedicate this book to him.

Leuven, April 2003 *Joris De Schutter*

Additional information, movies of the experiments, a camera calibration toolbox, the graphical user interface for contour fitting and some programming code are available on `http://www.khlim.be/~jbaeten/star.html`

Contents

1 Introduction 1
1.1 Approach 1
1.2 Main Contributions 2
1.3 Chapter by Chapter Overview 3

2 Literature Survey 5
2.1 Introduction 5
2.2 Feature Detection 6
2.2.1 Quality Criteria for Feature Detectors 6
2.2.2 Review of Edge Detection 6
2.2.3 Choices 8
2.3 Image Modelling 8
2.4 Visual Servoing 9
2.5 Camera Configuration 11
2.6 Hybrid Position/Force Control 11
2.7 Sensor Integration 12
2.8 Relation to This Work 14

3 Framework 17
3.1 Hybrid Control Structure 17
3.1.1 Example 19
3.1.2 Control Scheme 20
3.1.3 Robot with Joint Controller 20
3.2 Theoretical 2D Contour Following 23
3.2.1 Notations 23
3.2.2 Non-rotating Task Frame/Straight Contour 23
3.2.3 Rotating Task Frame/Straight Contour 25
3.2.4 Rotating Task Frame/Circular Contour 25
3.2.5 Conclusion 29

3.3 Dynamic Control Aspects ... 29
3.3.1 Vision Control Loop ... 30
3.3.2 Force Control Loop ... 31
3.4 Force Sensing ... 33
3.5 Tracking Error Identification ... 35
3.6 Relative Area Contour Detection ... 37
3.7 ISEF Edge Detection ... 40
3.8 Contour Modelling and Control Data Extraction ... 43
3.8.1 The Tangent Model ... 44
3.8.2 Comparison ... 45
3.8.3 Conclusion ... 48
3.9 Camera Model and Calibration ... 48
3.9.1 Used Camera Model ... 48
3.9.2 Calibration ... 49
3.9.3 Tsai's Model and Parameters ... 49
3.9.4 Tsai's Setup ... 50
3.9.5 Used Setup ... 51
3.9.6 Used Procedure ... 51
3.10 Conclusion ... 54

4 Classification ... 55
4.1 Introduction ... 55
4.2 Adopted Restrictions ... 56
4.3 Tool/Camera Configurations ... 58
4.4 Shared Control Types ... 61
4.5 Combined Vision/Force Task Examples ... 64
4.5.1 Traded Control Example ... 65
4.5.2 Hybrid and Shared Control Examples ... 66
4.6 Conclusion ... 71

5 Visual Servoing: A 3D Alignment Task ... 73
5.1 Strategy ... 73
5.1.1 3D Alignment Using a Circle ... 74
5.1.2 3D Alignment Using an Ellipse ... 77
5.1.3 3D Alignment Using a Rectangle ... 78
5.2 Used Algorithms ... 81
5.3 Experimental Results ... 82
5.3.1 Experimental Setup ... 82
5.3.2 Global Remarks ... 82
5.3.3 Circle Alignment Task ... 83
5.3.4 Ellipse Alignment Task ... 84
5.3.5 Rectangle Alignment Task ... 84
5.4 Conclusion ... 88

6 Planar Contour Following of Continuous Curves 89
6.1 Introduction 89
6.2 EOL Setup 90
6.3 Task Specification 91
6.4 Detailed Control Approach 92
6.4.1 Double Contact Control 92
6.4.2 Matching the Vision Data to the Task Frame 94
6.4.3 Contour Measurement 94
6.4.4 Tool Position Measurement 97
6.4.5 Matching 97
6.4.6 Calculating the Feedforward Signal 98
6.5 Experiments 100
6.5.1 Experimental Setup 100
6.5.2 Results 101
6.6 Conclusion 101

7 Planar Contour Following at Corners 105
7.1 Introduction 105
7.2 Path Measurement and Corner Detection 105
7.3 Augmented Control Structure 108
7.4 Experimental Results 110
7.5 Conclusion 112

8 Additional Experiments 115
8.1 Introduction 115
8.2 Traded Control in EOL 115
8.3 Shared Control in Non-parallel ECL 116
8.4 3D Path in Fixed EOL 118
8.5 Conclusion 119

9 Conclusion 123
9.1 Main Contributions 124
9.2 Limitations and Future Work 127

A Derivations 129
A.1 Frames and Homogeneous Transformations 129
A.2 Screw Transformations 130
A.3 Denavit-Hartengberg for KUKA361 131
A.3.1 Formalism 132
A.3.2 KUKA361 132
A.4 Point Symmetry with Relative Area Method 133
A.5 Relative Area Algorithms 136

A.6 Arc Following Simulation ... 137
A.6.1 The System ... 137
A.6.2 The Sensor ... 139
A.6.3 The Controller ... 140
A.6.4 Parameters ... 140
A.6.5 Some Results ... 140

B Contour Fitting ... 143
B.1 The Tangent Model ... 143
B.2 Full Second Order Function ... 144
B.3 Interpolating Polynomial ... 145
B.4 Third Order Non-interpolating Polynomial ... 145
B.5 Interpolating Cubic Splines ... 146
B.6 Parameterized Representation ... 149
B.7 Conclusion ... 151
B.8 Graphical User Interface ... 152

C Experimental Validation of Curvature Computation ... 155

D Image Processing Implementation on DSP ... 159

E Technical Documentation ... 161
E.1 Sinusoidal Contour ... 161
E.2 'Constant-Curved' Contour ... 162
E.3 Camera Mounting – 'Ahead' ... 162
E.4 Camera Mounting – 'Lateral' ... 164

References ... 167

Index ... 175

Symbols ... 177

1

Introduction

Sensor based robotics aims at the improvement of robustness, flexibility and performance of common robotic tasks. Integrated sensor based approaches can contribute to the ever growing industrial demand to increase velocity and precision of the robotic task, especially in an uncalibrated working environment. However, in contrast to the (highly appreciated) excellent kinematic capabilities of industrial robots, the potential of sensor based control in robotics is still underrated. Force and/or vision sensors, for example, are still an exception and mostly confined to the fields of scientific research.

Good sensing abilities are nevertheless essential to reach a higher level of flexibility and autonomy. Sensor based control can deal with and compensate for uncertainties in positions of workpiece and/or tool, e.g. due to tool wear or compliance. Also accurate fixture of the workpiece becomes superfluous.

Of all sensors, vision and force are among the most complementary ones. Visual sensors are powerful means for a robot to know its global task environment. Force sensors are essential in controlling the tool/workpiece contact or for monitoring interacting forces to protect the tool or the workpiece from damage. The *goal* of a mixed visual servoing/force control setup is to combine their advantages, while overcoming their shortcomings. Both the task quality, in the sense of increased velocity and accuracy, and the range of feasible tasks have to increase significantly with integrated vision/force control.

1.1 Approach

The presented, real-time approach aims at integrating visual servoing and force tracking based on the task frame formalism in a hybrid control structure. By dividing the control space into separate directions with clear physical meaning, the task frame formalism provides the means to easily model, implement and execute robotic servoing tasks in an uncalibrated workspace.

Distinct in our approach w.r.t. other hybrid vision/force research, is the *combined mounting* of vision and force sensor. Four meaningful force-tool/camera configurations are looked into. They are on the one hand the endpoint closed-loop configurations with either a parallel or a non-parallel camera mounting and on the other hand the endpoint open-loop configurations with either a fixed or a variable camera frame to task frame relation. Several tasks illustrate these configurations. Two concrete (variable EOL) applications are examined in detail: the 'planar contour following task' and the 'control approach at corners'. They show how the performance of a force controlled task improves by combining force control (tracking) and visual servoing. While maintaining the force controlled contact, the controller keeps the camera, also mounted on the robot end effector, over the contour at all times. Then, from the on-line vision-based local model of the contour, appropriate feedforward control is calculated and added to the feedback control in order to reduce tracking errors. The latter approach can be applied to all actions that scan surfaces along planar paths for one-off, uncalibrated tasks with a *rotationally symmetric* tool: cleaning, polishing, or even deburring ... .

1.2 Main Contributions

The main innovations and contributions of this book are:

- A new approach to *visual servoing* is explored by using the task frame formalism and relative area algorithms: Using the *task frame formalism*, the goal of the task is translated into control actions in the task frame. Each task frame directions is hereby linked to one specific image feature. All image features are computed using *relative area algorithms.*
- A new and **unique** framework for combined vision/force control is implemented. This framework is based on *the task frame formalism in a hybrid control setting* and uses *both sensors mounted on the end effector.*
  – The *task frame formalism* offers the means to easily model, implement and execute combined vision/force robotic servoing tasks in an uncalibrated workspace.
  – Vision and force sensing are combined in a *hybrid control scheme* by dividing the control space in **vision**, **force**, **tracking** and **velocity** subspaces, possibly with **feedforward**. The hybrid control scheme forms the heart of our approach and consists of a (sensor based) outer control loop around a (joint) velocity controlled robot.
- A new form of *shared* control in which vision based feedforward is added to force based tracking control, is presented together with examples of all possible shared control types.
- A new classification of (combined mounted) *camera/tool configurations* into parallel or non-parallel endpoint closed-loop and fixed or variable endpoint open-loop (EOL) is given. Numerous task examples illustrate these new configurations.

- Improved planar contour following for both continuous contours and contours with corners is realized using a variable endpoint open-loop setup and respective control strategy.
  – Adding vision based feedforward to the tracking direction improves the quality, in the sense of increased accuracy or execution velocity, of the force controlled planar contour following task at continuous curves or at corners.
  – A twice applied least squares solution is implemented for a robust, on-line computation of the curvature based feedforward.
  – For rotationally symmetric tools, there exists a redundancy for the rotation in the plane. This degree of freedom is used to position the camera while maintaining the force controlled contact. This results in a new *variable* relation between task and end effector frame orientations.
  – Special control issues arise, due to the time shift between the moment the contour is measured and the moment these data are used. A strategy to match the (vision) measurement data with the contact at hand based on Cartesian position is proposed.

1.3 Chapter by Chapter Overview

Figure 1.1 gives a structural overview of the following chapters. First, Chap. 2 briefly surveys the wide range of research domains which are related to our work. It places our approach in a global frame. Chapter 3 explains in more detail the different parts of the broad underlying structure that make up the framework of our approach. The heart of the framework is the hybrid control scheme. It forms the basis for the implementations in the next chapters. The separate elements of the hybrid control loop, including the sensor related issues, are discussed.

Chapter 4 builds on the presented framework by giving a comparative classification of combined vision/force tasks using different tool/camera configurations, on the one hand, and by exploring the possible types of shared control on the other hand. Numerous task examples (in 3D space) using combined visual servoing and force control illustrate the presented concepts. The remaining chapters investigate in more detail the key elements of these concepts.

First, to illustrate the visual servoing capabilities, Chap. 5 exemplifies a strategy to visually align the robot with end effector mounted camera relative to an arbitrarily placed object.

Then, Chaps. 6 to 8 study combined vision/force tasks. Especially, the endpoint open-loop configuration with variable task/camera frame relation, which is the most challenging in the sense of control, is fully explored in the planar contour following task of Chap. 6. Chapter 7 continues by extending the control approach to adequately round a corner. To end with, Chap. 8

presents the results of three additional experiments. They validate once more the potential of the used approach.

Finally, Chap. 9 concludes this work, followed by theoretical derivations, an evaluation of contour models, experimental validation of the curvature computation, some practical vision implementation issues and technical drawings with calibration results in appendices A to E respectively.

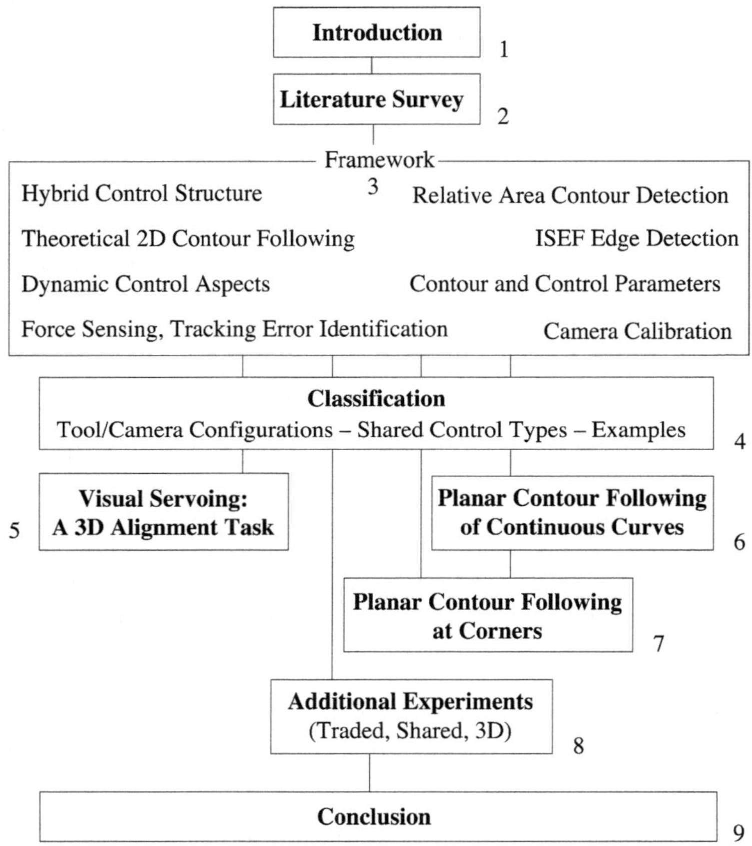

Fig. 1.1. Structural coherence of contents

2

Literature Survey

2.1 Introduction

This work spans and combines a wide range of research topics. Image processing, robot vision, image modelling, servo control, visual servoing, hybrid position/force control and sensor fusion are the most important ones. The following sections briefly review the relevant topics for each of these research domains and relate them to our work. Only stereo vision is left aside because of its fundamentally different nature.

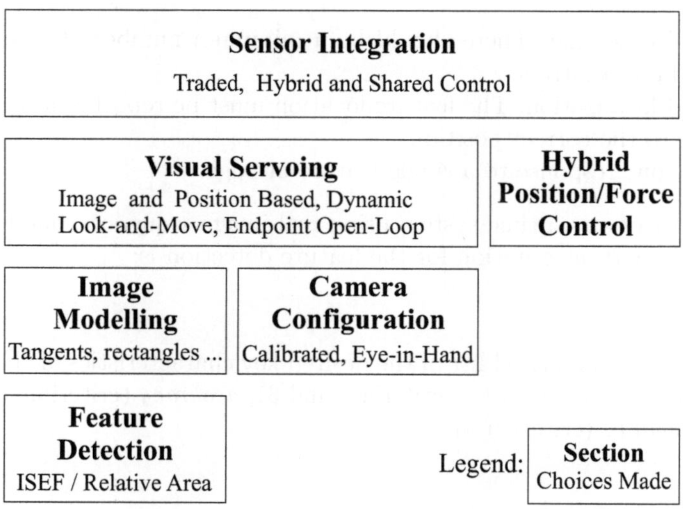

Fig. 2.1. Overview of treated topics in this section

Figure 2.1 gives an overview of the logical coherence of the treated topics. Section 2.2 starts with the criteria for and a review of the low level image processing. The suited techniques are chosen. Section 2.3 gives a brief overview of image modelling techniques, followed by an outline of the basic approaches in visual servoing in Sect. 2.4 and the camera configuration in Sect. 2.5. Then Sect. 2.6 briefly describes the standard issues in hybrid position/force control, the second pillar next to visual servoing, of this work. Section 2.7 continues with the fusion approaches for visual and force control. Finally, Sect. 2.8 summarizes our choices and approach.

2.2 Feature Detection

Image processing typically consists of three steps: (1) the preprocessing step, e.g. filtering and image enhancement, (2) the feature extraction, e.g. edge localization and (3) the interpretation, e.g. modelling, contour fitting or object matching. The low level image processing that implements the first two steps (often in one operation) is called *feature detection.* Typical image features are edges (one dimensional) and corners (two dimensional).

This section states the criteria for good feature detection, reviews the main feature detection techniques and reveals the choices made in this work.

2.2.1 Quality Criteria for Feature Detectors

Canny gives three criteria for feature (in particular edge) detection [12]. They are:

1. Good detection. There should be a minimum number of false negatives and false positives.
2. Good localization. The feature location must be reported as close as possible to the correct position.
3. Only one response to a single feature or edge.

Being part of a real time system using real image sequences, one final and at least as important criterion for the feature detection is:

4. Speed.

Another summarized list of the desired key characteristics of the detection algorithm is: *robustness* ($\sim$ criteria 1 and 3), *accuracy* (criterion 2) and *real-time feasibility* (criterion 4).

2.2.2 Review of Edge Detection

There has been an abundance of work on different approaches to the detection of one dimensional features in images. The wide interest is due to the large number of vision applications which use edges and lines as primitives, to achieve higher level goals. Some of the earliest methods of enhancing edges in images used small convolution masks to approximate the first derivative of the image brightness function, thus enhancing edges. These filters give very little control over smoothing and edge localization.

Canny described what has since become one of the most widely used edge finding algorithms [11]. The first step taken is the definition of the criteria which an edge detector must satisfy. These criteria are then developed quantitatively into a total error cost function. Variational calculus is applied to this cost function to find an "optimal" linear operator for convolution with the image. The optimal filter is shown to be a very close approximation to the first derivative of a Gaussian function.

Non-maximum suppression in a direction perpendicular to the edge is applied, to retain maxima in the image gradient. Finally, weak edges are removed using thresholding with hysteresis. The Gaussian convolution can be performed quickly because it is separable and a close approximation to it can be implemented recursively. However,the hysteresis stage slows the overall algorithm down considerably.

Shen and Castan propose an optimal linear operator for step edge detection [73]. They use similar analytical approaches to that of Canny, resulting in an efficient algorithm which has *exact recursive implementation.* The proposed algorithm combines an Infinite Symmetric Exponential smoothing Filter (ISEF), which efficiently suppresses noise, with a differential operator. It is shown that ISEF has a better performance in precision of edge localization, insensitivity to noise and computational complexity compared with Gaussian and Canny filters.

Venkatesh, Owens et. al. describe the approach of using "local energy" (in the frequency domain) to find features [86]. The local energy is found from quadrature pairs of image functions, such as the image function and its Hilbert transform. Fourier transforms of the image were initially used to find the required functions, at large computational cost.

Smith presents a new approach which involves a significant departure from feature extraction and noise reduction methods previously developed [76]. The brightness of each pixel within a mask is compared with the brightness of that mask's nucleus. The area of the mask which has the same (or similar) brightness as the nucleus is called the "USAN", an acronym for "Univalue Segment Assimilating Nucleus". From the size, centroid and second moments of the USAN two dimensional features and edges can be detected. Edges and corners correspond to the (locally) Smallest USAN value. This gives rise to the acronym SUSAN (Smallest Univalue Segment Assimilating Nucleus). This approach to feature detection has many differences with respect to the well known methods, the most obvious being that no image derivatives are used and that no noise reduction is needed.

2.2.3 Choices

In this work, two (different) low image processing techniques are investigated and used: a basic *pixel-weighting* or relative area method and the *ISEF edge detection* algorithm of [73].

Similar to the USAN method [76], the pixel-weighting method which directly determines centroids, distances and angles, described in Chap. 3 Sect. 3.6 and appendix A, has the advantage of being *very robust*. The integrating effect of the principle ensures strong noise rejection. This method is also *fast* because each pixel is evaluated only once and it is well suited for image-based visual servoing (in which feature characteristics are directly used as control inputs, see Sect. 2.4). The method has, however, *limited accuracy*, uses binary thresholding and is only applicable to local basic images.

To overcome the limited accuracy of the pixel-weighting method, ISEF edge detection is used. This optimal linear step edge detector is still quite *robust*, has *excellent accuracy* and is *fast*. The computational burden is especially low when the edge detector is used on small sub-images with only a limited number of line convolutions. The ISEF edge detector thus fulfills all the quality criteria expectations (Sect. 2.2).

2.3 Image Modelling

The next step, after feature detection, is the image/feature interpretation. This step models the features in the image, in order to generate (in our case) the control actions to be taken. Typically, an image feature will correspond to the projection of a physical feature of some object on the image plane. The modelling step then extracts the *image feature parameters* by using either *implicit* or *explicit* models as a representation of the physical object or target.

Implicit modelling assumes that the model is incorporated in the detection algorithm itself. For example, the proposed pixel-weighting method (Sect. 3.6) uses as implicit model a straight contour. The measurements of the distance to the contour and the angle of the contour are directly incorporated in the method. Implicit modelling is typically part of an image-based approach (see Sect. 2.4).

Most techniques, however, use explicit models with feature parameters such as (relative) distance, (relative) area, centroid and higher order moments. For example, Chaumette, Rives and Espiau [14, 34] use a set of (four) points or lines for relative positioning and visual servoing; Sato and Aggarwal [71] estimate the relative position and orientation of a camera to a circle having unknown radius; Jörg, Hirzinger et al. [47] also use a circle in tracking an engine block; and Ellis et. al. [33] consider the fitting of ellipses to edge-based pixel data in the presence of uncertainty. More complex object models, made possible by the emerged computational power, are e.g. used in robust workpiece tracking and gripping by Tonko et al. [79]; in real-time tracking of 3D structures by Drummond and Cipolla [32] or by Wunsch and Hirzinger [94].

Typical contour models are (a set of) lines [22,23], tangents [9,54], a second order curve [54], a polar description [16] and splines [54] or snakes e.g. [45]. Chapter 3 investigates different contour models for vision based contour following going from full second order curves (circle, ellipse, hyperbole ...), over splines to a list of points and tangents. Circle models are found to be appropriate if the contour is (in advance) known to be circular, otherwise a list of points and tangents, which is used in our approach, gives better results.

The usefulness of a particular model not only depends on the robustness, accuracy and speed in fitting the model onto the edge data but also on the robustness, accuracy and speed at which the control signals are derived from the model or thus from the image feature parameters. Robustness, e.g. in object recognition and classification or even visual servoing, can be achieved by the use of *invariant* image feature parameters in the representation model [51, 82, 84].

The counterpart of the feature-based methods, discussed until now, is the *correlation-based* method. If the target has a specific pattern that changes little over time (i.e. temporal consistency), then tracking can be based on correlating the appearance of the target in a series of images. The Sum of Squared Differences (SSD) [65, 66, 75] is such a correlation-based method[1] which can be used to determine the relative motion between object and camera, to get depth or structure from motion in mono-vision or even to initialize a set of good trackable features. On the other hand, if the target does not show a specific pattern, as is the case e.g. in an image of a straight contour, a correlation-based method will result in multiple inconclusive correspondences between the original and shifted images. Hence, it is unfit for the contour following tasks at hand. A technique related to the correlation-based methods, but much more sophisticated, is optical flow [41].

2.4 Visual Servoing

Sanderson and Weiss [70, 92] introduced a taxonomy of visual servo systems, into which all subsequent visual servo systems can be categorized [19, 36]. Their scheme distinguishes dynamic look-and-move from direct visual servo systems on the one hand and image-based from position-based ones on the other.

Dynamic look-and-move versus direct visual servo

If the vision system provides set-point inputs to the joint-level controller,

[1] For the correlation methods too, either implicit models, e.g. in feature initialization, or explicit models, e.g. in pattern recognition, are used

which internally stabilize the robot, it is referred to as a *dynamic look-and-move* system. In contrast, *direct visual servo*[2] eliminates the robot controller entirely replacing it with a visual servo controller that directly computes joint inputs, thus using only vision to stabilize the mechanism. Most implemented systems adopt the dynamic look-and-move approach for several reasons [44]. First, the relatively low sampling rates available from vision[3] make direct control of a robot end effector with complex, nonlinear dynamics an extremely challenging problem. Second, many robots already have an interface for accepting Cartesian velocity or incremental position commands. This simplifies the construction of the visual servo system, and also makes the methods more portable. Thirdly, look-and-move separates the kinematic singularities of the mechanism from the visual controller, allowing the robot to be considered as an ideal Cartesian motion device. This work utilizes the look-and-move model only.

Position-based visual servoing versus image-based visual servoing
In *position-based visual servoing* (PBVS) (e.g. [57]), the image modelling step reconstructs (with known camera model) the (3D) workspace of the robot. The feedback then reduces the estimated pose errors in Cartesian space. If a geometric object model lies at the basis of the Cartesian pose estimation, the approach is referred to as *pose-based.*
In *image-based visual servoing* (IBVS) (e.g. [16,35,38,48]), control values are computed from the image features directly. The image-based approach may eliminate the necessity for image interpretation and avoids errors caused by sensor modelling and camera calibration. It does however present a significant challenge to controller design since the system is non-linear and highly coupled.

In this context, the *image Jacobian* was first introduced by Weiss et al. [92], who referred to it as the feature sensitivity matrix. The image Jacobian is a linear transformation that maps end effector velocity to image feature velocities. Examples can be found in [34,38,44] or in [2,42]. In their work Asada et al. [2,42] propose an adaptive visual servoing scheme with on-line estimation of the image Jacobian. McMurray et al. [68] even incorporate the unknown robot kinematics in the dynamic on-line estimation of the Jacobian.

Endpoint open-loop versus endpoint closed-loop
In addition to these considerations, a distinction is made between systems which only observe the target object, as is mostly the case in this work, and those which observe both target object and robot end effector [19,44]. The

[2] Sanderson and Weiss originally used the term "visual servo" for this type of system, but since then this term has come to be accepted as a generic description for any type of visual control of a robotic system.

[3] In spite of reports on 1 Khz visual servoing [46,58] most visual servoing researchers still prefer to use the low cost standard rate vision systems.

former are referred to as *endpoint open-loop* (EOL) systems, and the latter as *endpoint closed-loop* (ECL) systems. The primary difference is that EOL systems must rely on an explicit hand-eye calibration when translating a task specification into a visual servoing algorithm, especially in the position-based visual servoing case. Hence, the positioning accuracy of EOL systems depends directly on the accuracy of the hand-eye calibration. Conversely, hand-eye calibration errors do not influence the accuracy of ECL systems [1, 93].

2.5 Camera Configuration

Visual servo systems typically use one of two camera configurations: *end effector mounted*, or *workspace related*[4]. For the first, also called an eye-in-hand configuration, there exists a known, often fixed, relationship between the camera pose (position and orientation) and the end effector pose. In the workspace related configuration, the camera pose is related to the base coordinate system of the robot. Here, the camera may be either fixed in the workspace or mounted on another robot or pan/tilt head. The latter is called an *active camera*[5] system.

For either choice of camera configuration, some form of camera calibration must be performed. In this work, Tsai's calibration technique [52, 80, 81] is adopted.

Chu and Chung [15] propose a selection method for the optimal camera position in an active camera system using visibility and manipulability constraints. Also in this context, Nelson and Khosla [60, 62] introduced the vision resolvability concept: the ability to provide good feature sensitivity, a concept closely related to the image Jacobian.

In this work, we choose to mount the camera on the end effector. This results in local images and avoids occlusion. Moreover, it gives a controllable camera position to the effect that the image feature of interest can be placed on the optical axis (center of the image) and tracked by visual servoing. Hence, the feature measurement is less sensitive to calibration errors or distortions in the camera/lens system. Furthermore, placing the camera close to the target ensures a good resolution in the image feature measurement.

2.6 Hybrid Position/Force Control

The two basic approaches to force controlled manipulators [13, 74], surveyed by Yoshikawa in [98], are: *hybrid position/force control* and *impedance control*. The impedance control approach proposed by Hogan [40] aims at controlling

[4] Note that a mounted camera does not necessarily imply an EOL system. Equally, a fixed camera configuration is not by definition an ECL system.

[5] This subject is closely related to active sensing research.

position and force by translating a task into a desired impedance, a relation between motion and force. The hybrid control approach was originally proposed by Raibert and Craig [69]. It separates the control space in position and force controlled directions, resulting in two perpendicular subspaces: the (position or) velocity controlled subspace and the force regulated subspace also called twist and wrench spaces respectively[6].

In case of a point contact between tool and workpiece, a hybrid position force controlled task can easily be defined in the task frame formalism [10, 26, 28, 55]. Here, desired actions are specified separately for each direction of an orthogonal frame attached to the tool: *the "Task Frame"*. Each direction (x,y,z) is considered once as an axial direction (translation; linear force) and once as a polar direction (rotation; torque).

This work is based on the hybrid position/force control approach and on the task frame formalism as adopted, among others [39], by De Schutter and Van Brussel [10, 27, 28].

2.7 Sensor Integration

The last step, of utmost importance to this work, is the sensor integration, which combines visual servoing and force control. Only recently, researchers have started to merge the fundamentals of both visual servoing and force controlled robotics, in order to benefit from the complementary characteristics of vision and force sensors. A combined vision/force setup, made possible by the emerged computational power and low cost vision systems, expands the range of feasible robotic tasks significantly. The two sensing systems, however, produce fundamentally different quantities, force and position. This may present an inherent problem when attempting to integrate information from the two sensors. Sensor integration techniques require a common representation among the various sensory data being integrated. Force and vision sensors do not provide this common data representation. Furthermore, as a consequence of their different nature, the two sensors are often only useful during different stages of the task being executed. Hence, force and vision information is seldom fused as redundant data [60, 97]. We do however present tasks in which vision and force sensing is really fused. This is for example the case when vision based feedforward is added to force related feedback.

Nelson et al. [61] presented three basic strategies which combine force and vision within the feedback loop of a manipulator: traded control, hybrid control and shared control. In *traded control*, the controller switches between force control and visual servoing for a given direction; *Hybrid control*, an extension to the previously mentioned hybrid position/force control, allows visual servoing only in the twist space; The highest level of integration however

[6] The orthogonality of twist and wrench spaces can be questioned. For the tasks at hand, however, this concept stands.

involves *shared control*: it implies the use of visual servoing and force control along the same direction simultaneously. Our approach includes hybrid, traded as well as shared control.

Another approach, which is not adopted in this work but somehow related to vision based feedforward on a force controlled direction, is given by Morel et al. [56]. They propose an impedance based combination of visual and force control. In their scheme, the vision control loop generates the reference trajectory for a position based impedance controller with force feedback. Thus, the visual control is built around the force control loop, in contrast to the previously presented basic strategies [61] where vision and force control act at the same hierarchical level.

Some researchers focus on hybrid vision/force control in a fully uncalibrated workspace. E.g. [43, 67, 97] use fixed cameras which are not calibrated w.r.t. the robot. Their approaches require on-line identification of the contact situation, separation of the workspace in force and vision controlled directions, on-line vision Jacobian identification and adaptive control, each of which is tackled successfully. However, defining a path in the image plane of an uncalibrated camera, has no known physical meaning. We will therefore only assume the object space to be uncalibrated. Hence, only position, orientation and shape of the workpiece or object are unknown[7]. Any other information about the task is translated in a *high level task description* which defines the complete task, divided in sub-tasks, using the task frame formalism. The task frame formalism thus provides the basis to combine vision and force sensing in an easy but powerful and flexible way with clear physical meaning. The high level task description divides the six degree of freedom (DOF) cartesian control space in orthogonal, separate or mixed, vision and force (among others) controlled directions, depending on the type of task or expected contact at hand. This, in contrast to [43, 67, 95–97], facilitates the decoupling of vision and force control subspaces, if needed. Hosoda et al. [43] also suggest the need for a common coordinate frame for multiple external sensor-based controllers. In our opinion, the task frame lends itself very well to this purpose for numerous tasks, especially when multiple sensors are mounted on the end effector. Furthermore, since the high level task description determines the use of each sensor, it counters the need for vision and force resolvability as introduced by Nelson [60].

For the combined vision/force contour following, as presented in this work, the camera needs to be calibrated with respect to the robot end effector. After all, with the camera looking ahead and not seeing the actual contact point (i.e. endpoint open-loop), calibration is essential to match the vision information to the contact situation at hand. Thanks to the calibration, the vision algorithms too will equally benefit from the combined vision/force setup. If the relative position of camera and force sensor is known and if the (calibrated) tool is in

[7] This is especially true for one-off tasks in which accurate positioning or calibration of the workpiece is costly or impossible, causing large pose errors.

contact with the object, the depth from the image plane to the object plane is known. Hence, the 3D position of the image feature can be calculated easily with mono vision. This is yet another type of sensor integration.

Other reports on combined vision/force applications or approaches are given by Zhou, Nelson and Vikramaditya [63, 99] in micromanipulation and microassembly, by von Collani et al. [88–90] in a neuro fuzzy solution for integrated visual and force control, by Song et al. [77] in a shared control structure, including vision and force sensing, for an arm/hand teleoperated system, by Hirzinger et al. [47] in a flexible robot-assembly using a multi-sensory approach, by Malis, Morel and Chaumette [53] in vision/force integration using the task function approach and by Namiki et al. [59] in high speed grasping.

2.8 Relation to This Work

This work presents two approaches. They can be categorized as follows. Chapter 5 gives a *3D relative positioning task*

- by image based visual servoing
- using pixel-weighting or relative area related methods,
- with a coarsely known camera setup.

Chapters 6 to 8 present *combined vision/force contour following tasks*[8]

- with calibrated camera
- using the ISEF edge detection
- with on-line contour modelling by tangent lines
- for position based visual servoing
- in a hybrid position/force control scheme
- including hybrid, traded as well as shared control.

All the presented methods are:

- feature based,
- dynamic look-and-move,
- (mostly) endpoint open-loop,
- eye-in-hand and
- based on the task frame formalism.

Unlike most reported hybrid position/force control with force sensor and camera [43, 59–62, 64, 67, 95–97], our work uses an *eye-in-hand camera*, rather than a fixed one. Few researchers reported combined vision force control with both sensors mounted on the manipulator. Nevertheless, they mostly use completely different control approaches, being neuro-fuzzy [88] or impedance based [56]. Only recently, Jörg, Hirzinger et al. [47] and Malis, Morel and Chaumette [53] have published multiple sensor control approaches similar to

[8] See also [3–9]

our. The former work uses multiple sensors in a flexible robot-assembly. The chosen task, however, lends itself only for traded control. The latter uses, in contrast to the task frame formalism, a task function approach, which they introduced in earlier publications to define the goal in a visual servoing task and now extended to include force control.

The end effector mounted camera in an endpoint open-loop control enforces the need for camera calibration (pose and intrinsic camera parameters). The workpiece and environment on the other hand may be fully uncalibrated.

The task frame formalism is used as a concept to combine vision and force control in an easy but powerful and flexible way with clear physical meaning. In addition, vision based feedforward is added. The use of the task frame formalism results by definition in orthogonal, separate or mixed, force and vision control spaces. This avoids the extra effort of decoupling these spaces. It also distinguishes our work from visual servoing such as [34, 44]: neither a vision Jacobian identification nor optical flow calculations are necessary.

Many researchers report on vision based feedforward control. In contrast to this work, however, they mostly use off-line generated models (e.g. [25]) or partially known paths and workpieces (e.g. [50]).

3

Framework

The previous chapter gave an overview of combined visual servoing and force control in robotics together with some basic underlying topics. This chapter will explain in more detail the different parts of the broad underlying structure that make up the framework of our approach as a basis for the next chapters.

The first section discusses the hybrid control structure used in our approach. It divides a robotic task into separately controlled directions using the task frame formalism. The overall control architecture consists mainly of three parts: the controller, the system and the sensing modalities. The former two are briefly discussed in the last part of the first section. The sensors are the subject of Sects. 3.4 to 3.9.

After the introduction of the hybrid control scheme in Sect. 3.1, Sect. 3.2 gives the theoretical foundation for the fitness of the chosen control structure. Together with Sect. 3.3, it surveys the static and dynamic characteristics of vision and force control loops.

The subsequent sections treat the remaining approach issues involved in the used sensing techniques. First, Sect. 3.4 briefly treats the force sensor and its characteristics. Sections 3.5 explains the tracking error identification based on either measured forces or velocities. Sections 3.6 and 3.7 describe the adopted low level image processing techniques, followed by a comparative discussion of the properties of the chosen contour model in Sect. 3.8. Finally, Sect. 3.9 describes the projective camera model and explains the used camera calibration procedure. Section 3.10 concludes the chapter.

3.1 Hybrid Control Structure

The used hybrid control structure combines sensor based position control with velocity set-points, implemented as an outer control loop around the joint controlled robot [27]. The basis for a common representation in this hybrid control is the task frame formalism [10, 28, 55]. In this formalism, desired actions are specified separately for each direction of an orthogonal frame related to the task: *the Task Frame* (TF). Each direction (x,y,z) is considered once as an axial direction and once as a polar direction. Force, position and vision sensors can each command separate TF directions. In total, we distinguish five types of directions: velocity, force, vision, tracking and feedforward.

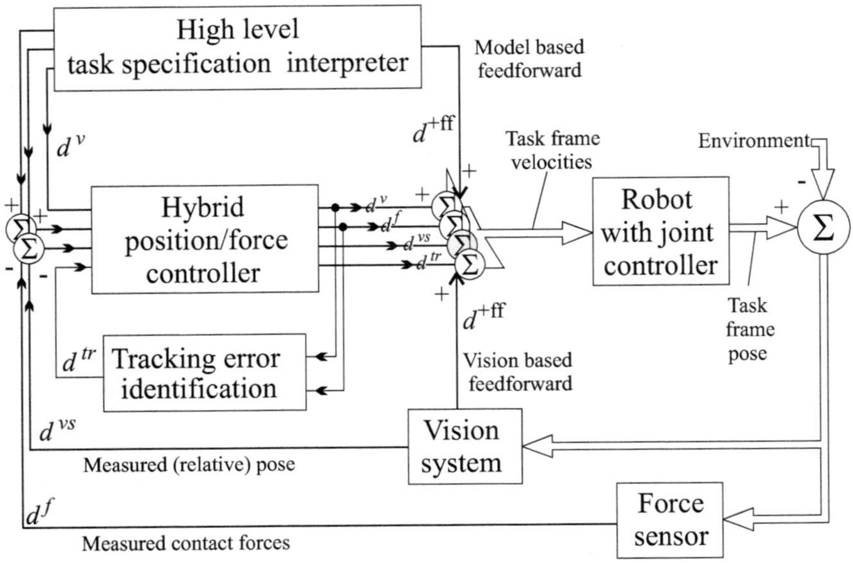

Fig. 3.1. Hybrid control scheme

Velocity direction: If the robot is commanded to move at a desired velocity in a given direction of the task frame, we call this direction a *velocity direction.* Normally, there are no (physical) motion constraints[1] in a velocity direction. The desired velocity is specified in the task description. The 'set-point' direction d^v of Fig. 3.1 is a velocity direction.
Force direction: If the desired action in a given TF direction consists of establishing and maintaining a specified contact force with the environment or object, we call this direction a *force controlled direction* or in short a *force direction.* Control loop d^f in Fig. 3.1 controls the contact forces in a force direction. Normally, in a force direction, the TF interacts with the environment. This does however not always need to be so. For example, the first stage of a guarded approach in a force controlled direction, can be a free space motion.
Vision direction: The TF direction used to position the task frame relative to the selected image features by visual servoing, is a *vision controlled direction* or in short a *vision direction.* Control loop d^{vs} in Fig. 3.1 is vision controlled. Normally, there are no motion constraints[1] in a vision controlled

[1] imposed by the environment onto the task frame

direction. The vision measurements are in se pose errors[2] (of the object w.r.t. the task frame). Hence, a vision direction is in fact a position controlled direction using a vision sensor. A vision direction shows similarities to a force direction, since both are sensor controlled, although with quite different sensor characteristics.

Tracking direction: If the controller uses a given TF direction to automatically adjust the pose of the *moving* task frame w.r.t. the environment, we call this direction a *tracking direction*. For a tracking direction, no desired velocity is specified. Implicitly, a tracking error equal to zero is asked. Control loop d^{tr} of Fig. 3.1 tries to minimize the (absolute value of the) tracking error in a tracking direction. Since the basis for the tracking control is a positional error, a tracking direction shows similarities to a vision direction. However, the identification of the tracking error in a tracking direction is fundamentally different from the the tracking error measurement in a vision direction. Section 3.5 discusses the identification of the tracking error.

Feedforward direction: Any direction of the task frame, whether it is a velocity, a vision, a force or a tracking direction, to which a feedforward signal is added, is (in addition) called a *feedforward direction*, indicated by $d^{+\text{ff}}$ in Fig. 3.1. The feedforward signal is superimposed on the existing velocity signals and is either model based or sensor based [24,25] (in casu vision based). The objective of a feedforward action is to reduce errors in a tracking, force or vision direction. Section 3.2 gives a theoretical foundation for the needed feedforward signal in planar contour following.

Note that *position control*, without the use of external sensors, is not possible with the given hybrid control scheme. If necessary, position control of the robot end effector is performed by another (separate) controller, which is not discussed. The integrating of position control with the described velocity based controller, such that the joint controller gets a mix of desired TF velocities and positions as input, forms a great challenge for future controller design[3].

3.1.1 Example

In the case of 3 DOF (Degrees Of Freedom) planar (compliant) motion as shown in Fig. 3.2, the task is specified by the desired tangential velocity (direction type d^v) and the desired normal contact force (direction type d^f). Furthermore, the plane normal is used as a polar tracking direction (d^{tr}) to keep the task frame tangent to the contour. All other TF directions are in fact velocity directions (d^v) with specified velocities set equal to zero. To the tracking direction feedforward may be added ($d^{+\text{ff}}$). The better the feedforward velocity anticipates the motion of the task frame, the smaller the tracking

[2] pose: position and orientation

[3] Only recently, Lange and Hirzinger [49] reported on such a mixed position/velocity integrated control structure.

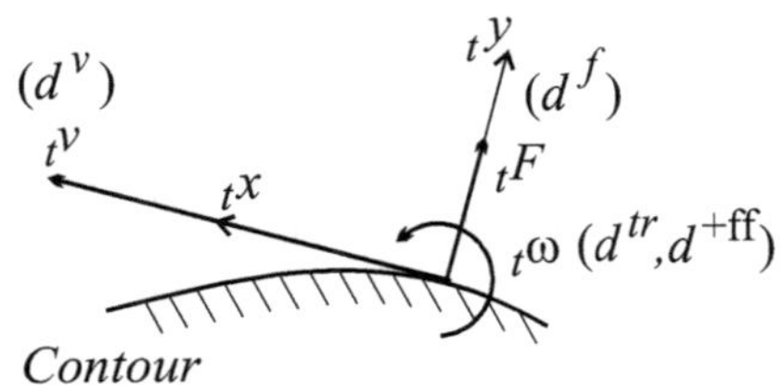

Fig. 3.2. 3 DOF planar compliant motion (with the control type for a given direction indicated between brackets)

error. Hence, the faster the task can be executed before the errors become so large that contact is lost or excessive contact forces occur [27,28] or the better the contact forces are maintained.

Figure 3.3 gives a typical setup. The (6DOF) Cartesian pose of any part of the setup is defined by orthogonal frames attached to the considered part. Aside from the task frame, the figure shows the camera frame, the force sensor frame, the end effector frame and the absolute world frame. The task frame is denoted by (preceding) subscript t, the other frames by subscripts cam, fs, ee and abs respectively.

3.1.2 Control Scheme

The hybrid control scheme of Fig. 3.1 forms the basis of our approach. The hybrid controller implements proportional control laws for each controlled direction (vision, force or tracking). The output of the controller are commanded velocities. For a velocity direction the commanded velocities are equal to the specified (desired) velocities. Table 3.1 gives an overview of the control actions for velocity, force, vision and tracking directions, in either axial or polar cases. To each velocity signal feedforward can be added. Sections 3.2 and 3.3 discuss the control loop characteristics. They demonstrate the fitness of the chosen proportional control laws with feedforward.

Together, all the commanded velocities, grouped in a 6×1 Cartesian TF velocity vector or twist, (must) define unambiguously the instantaneous motion of the task frame in 6 DOF Cartesian space. This velocity vector is the input for the joint controlled robot.

3.1.3 Robot with Joint Controller

For the sake of completeness, Fig. 3.4 shows the inner servo control loop enclosed by the "robot with joint controller"-block of Fig. 3.1. The TF velocity vector is first recalculated[4] to absolute end effector velocities and then transformed into the desired joint velocities, using the inverse robot Jacobian for

[4] The velocity screw transformation S^v is formulated in Appendix A.

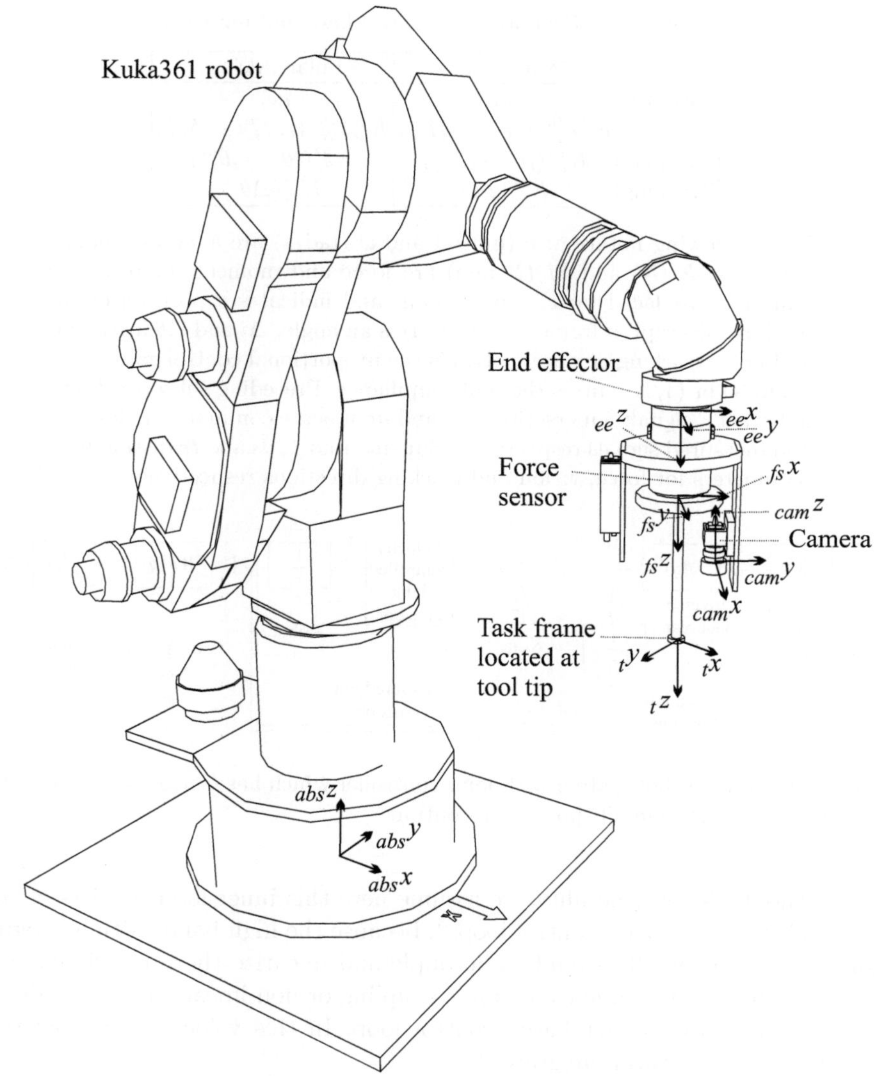

Fig. 3.3. Typical setup with definition of task, end effector, camera, force sensor and absolute frames

the current robot position. The joint velocities determine the desired joint positions and are also used as feedforward velocities in the joint control loops for each axis of the robot. The final output of the robot is the absolute world Cartesian pose of the task frame (based on the physical forward kinematic structure of the robot end effector and the relation of the task frame to the robot end effector).

Table 3.1. Proportional control laws and notation

	Axial : ${}_tv_i^c =$	Polar : ${}_t\omega_i^c =$
Velocity	${}_tv_i^d$	${}_t\omega_i^d$
Force	$K_i^f k_i^{inv}({}_tF_i^m - {}_tF_i^d)$	$K_{\theta i}^f k_{\theta i}^{inv}({}_tM_i^m - {}_tM_i^d)$
Vision	$K_i^{vs}({}_ti^d - {}_ti^m))$	$K_{\theta i}^{vs}({}_t\theta_i^d - {}_t\theta_i^m)$
Tracking	$K_i^{tr}(\Delta i)$	$K_{\theta i}^{tr}(\Delta\theta_i)$

With following notation: v (mm/s) and ω (rad/s) are axial and polar velocities; F (N) and M (N mm) are force and moment (or torque); i must be replaced by x, y or z (mm) and indicates a position or, if used as subscript, a frame axis; θ (rad) is an angle; Δi and $\Delta\theta$ are axial and polar tracking errors; K (1/s) is the proportional control gain; k^{inv} (mm/N) or (1/N mm) is the tool compliance; Preceding subscript t denotes a TF signal; Superscripts c, d and m indicate commanded, desired and measured signals respectively; Superscripts f, vs and tr distinguish parameters for force, vision and tracking directions respectively.

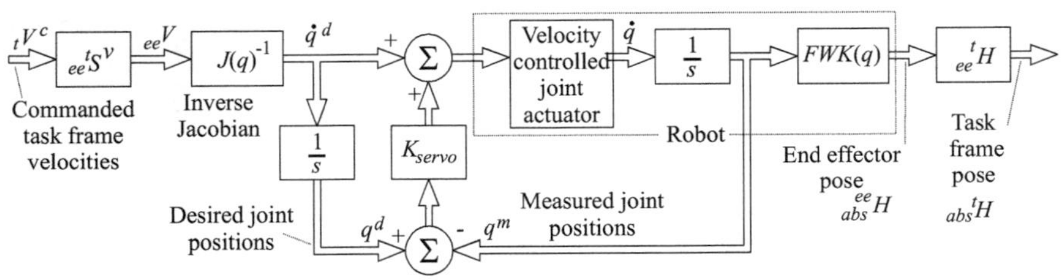

Fig. 3.4. Actual robot system with joint controller which has Cartesian TF velocity as input and Cartesian TF position as output

Without loss of generality, we assume here this inner loop consists of *independent* joint velocity control loops[5], because the high bandwidth analogue joint velocity controllers tend to decouple and linearize the robot dynamics. Any remaining errors due to dynamic coupling or non-lineair effects are dealt with by the outer sensor based control loop. In this velocity approach the robot acts as a natural integrator[6].

The joint velocity feedforward ($\dot{q}^d$) improves the dynamical behaviour of the robot. If an industrial position controlled robot were to be used, this would not be possible and the input to the robot would be the desired joint positions only (q^d).

The remaining parts of the control block diagram of Fig. 3.1, being the force sensor, the tracking error identification and the vision system, are discussed in more detail in the Sect. 3.4 to 3.9.

[5] We can always use a more complex controller based on a dynamic model of the robot

[6] The transfer function from desired joint velocities ($\dot{q}^d$) to actual joint positions (q) can be approximated by a set of decoupled pure integrators.

3.2 Theoretical 2D Contour Following

This section describes the control approach and steady state errors involved in following a 2D planar contour. Three situations are examined: (1) contour following of a straight line, which lies at a fixed angle w.r.t. the forward direction of the none-rotating task frame, (2) contour following of a straight line with a rotating task frame and (3) contour following of an arc shaped contour with a rotating task frame.

The given analysis is similar to the one given by De Schutter [30,31] which applies to compliant motion with force control in a direction normal to the path.

3.2.1 Notations

Following notations are used. ${}_{abs}^{\ t}x$ is the absolute x-position of the task frame (mm); ${}_{abs}^{\ t}\theta$ is the absolute orientation of the task frame (rad). ${}_{t}^{c}x$ and ${}_{t}^{c}\theta$ are the relative position and orientation of the contour w.r.t. the task frame. For reasons of simplicity, the preceding superscript c is sometimes omitted. Additional superscripts c, d and m indicate commanded, desired and measured signals respectively; Superscript ff denotes a feedforward signal; superscript ss indicates a steady state signal. K_x and K_θ (1/s) are the proportional gains for the ${}_tx$- and ${}_t\theta$-control loops respectively. In each case the TF moves at a constant velocity ${}_tv_y^c$ in the ${}_ty$-direction[7].

3.2.2 Non-rotating Task Frame/Straight Contour

Following a straight contour with a non-rotating task frame aims at minimizing the difference in the ${}_tx$-direction between the origin of the task frame and the contour while moving with a fixed velocity in the ${}_ty$-direction of the task frame. Figure 3.5 gives the simplified proportional position control loop for the ${}_tx$-direction.

The commanded ${}_tx$-velocity ${}_tv_x^c$, according to a proportional control law, is equal to $K_x({}_tx^d - {}_tx^m)$. If the robot moves immediately at the commanded velocity, the relation between the output of the robot, being a position, and the input, being a velocity, is a simple integrator. Hence, the simplified transfer function of the robot is $R(s) = 1/s$, with s the Laplace variable.

Following a straight line, at a fixed angle ${}_{t}^{c}\theta$ with the forward ${}_ty$-direction of the non-rotating task frame, as illustrated in Fig. 3.6, causes a (constant) steady state tracking error, defined by

$$\Delta x^{ss} \equiv \lim_{t\to\infty} ({}_tx^d(t) - {}_tx^m(t)). \tag{3.1}$$

[7] Hence, there is no (external sensor) control for the position of the TF in the ${}_ty$-direction.

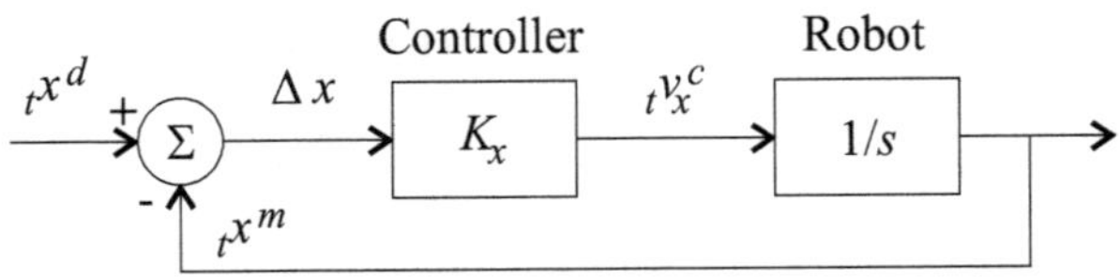

Fig. 3.5. Simplified (one-dimensional) position control loop

According to the final value theorem of Laplace, this steady state tracking error equals[8]

$$\Delta x^{\text{ss}} = \lim_{s\to 0} s \left(\frac{s}{s+K_x} \cdot \frac{-_t v_y \tan(^C_t\theta)}{s^2} \right) \tag{3.2}$$

$$= \frac{-_t v_y \tan(^C_t\theta)}{K_x}, \tag{3.3}$$

with $s/(s+K_x)$ the closed loop transfer function from input to $_tx$-direction error and $-_tv_y \tan(^C_t\theta)/s^2$ the input signal. Equation 3.2 implies that in steady state the task frame moves parallel to the desired contour[9].

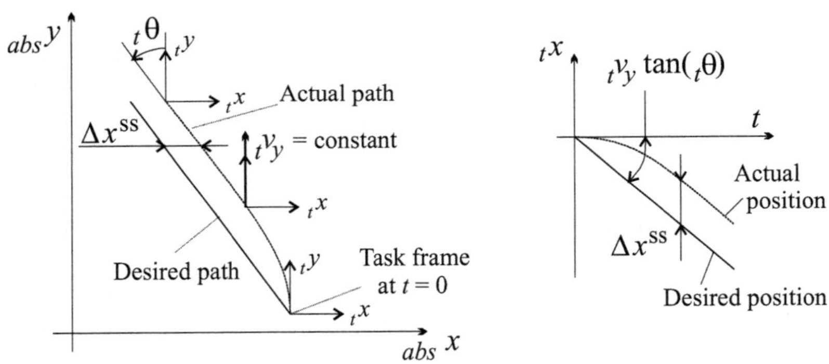

Fig. 3.6. 2D path (**left**) and time response (**right**) for the proportional $_tx$-direction control of a non-rotating task frame following a straight contour at a fixed angle.

As is well known from control theory, the steady state tracking error will vanish for the given setup, if the controller itself, aside from the robot being an integrator, also contains an integrator. Adding an extra integrator, however, may endanger the stability of the control loop. A better solution to eliminate the tracking error, not affecting the stability, is the use of feedforward control. The additional (tangent based) feedforward component

8 This equation corresponds to the steady state force error ΔF derived in [30] by substituting Δx by ΔF and K_x by $K^f.k^{inv}$, which are the force control gain and compliance.

9 If we assume this parallel path to be valid in the first place, so $_tv_x^{\text{ss}} = -_tv_y \tan(^C_t\theta)$, and apply the control law to the steady state equilibrium, giving $_tv_x^{ss} = K_x.\Delta x^{\text{ss}}$, then the result of (3.2) directly follows from these latter two equations.

$$ {}_t v_x^{\mathrm{ff}} = {}_t v_y \tan({}^c_t\theta), \tag{3.4}$$

as shown in Fig. 3.7, keeps the center of the task frame on the desired path, once it has reached this desired path. Any initial error in the task frame position is eliminated by the feedback control. Hence, the tracking error becomes

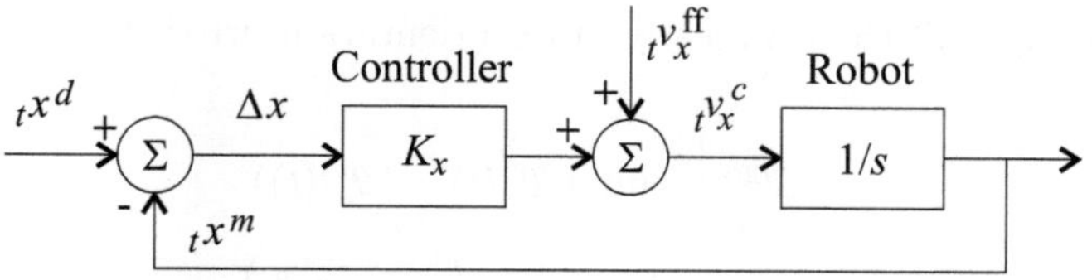

Fig. 3.7. One dimensional position control loop with feedforward

zero with (tangent based) feedforward. Equation (3.4) implies that the angle ${}^c_t\theta$ of the contour is measurable. With vision, but also by force tracking, this is the case.

3.2.3 Rotating Task Frame/Straight Contour

A last way to eliminate the tracking error in following a straight line, is the use of a rotating instead of a non-rotating task frame. This fundamentally changes the setup. Here, the controller regulates both the position (in the ${}_tx$-direction) and the orientation (about the ${}_tz$-direction) of the task frame, while moving with a fixed velocity in the ${}_ty$-direction of the task frame. Hence, the ${}_tx$-position control loop of Fig. 3.5 is augmented with a ${}_t\theta$-orientation control, given in Fig. 3.8.

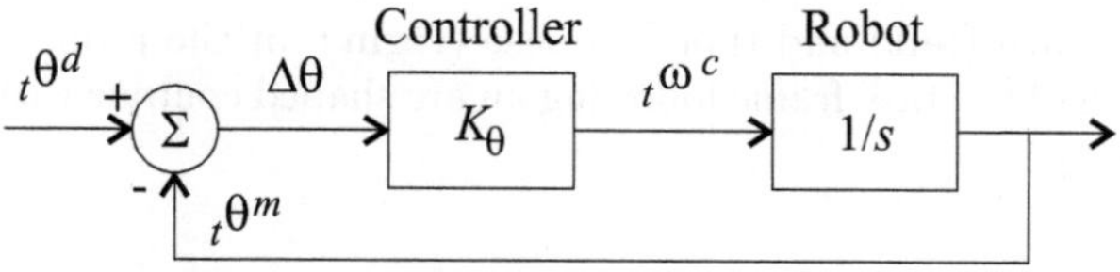

Fig. 3.8. One dimensional control loop for the TF orientation

Controlling the orientation of the task frame changes the forward direction, i.e. the ${}_ty$-direction, in time, hereby decreasing the angle ${}^c_t\theta$. When ${}^c_t\theta$ becomes zero, the task frame no longer rotates and the steady state tracking error in the ${}_tx$-direction will also become zero. This follows from (3.2) with ${}^c_t\theta = 0$.

3.2.4 Rotating Task Frame/Circular Contour

As previously mentioned, the 2D control of a rotating task frame contains two control loops: the ${}_tx$-position control loop of Fig. 3.5 with proportional control constant K_x and the ${}_t\theta$-orientation control loop of Fig. 3.8 with proportional control constant K_θ. The ${}_ty$-velocity is fixed. The control objective is to move along the contour with constant ${}_ty$-velocity while keeping the ${}_ty$-direction tangent to the contour and to minimize the error in ${}_tx$-direction.

If the path to be followed is arc shaped (with radius r) or, in other words, if the desired orientation of the task frame increases linearly with time, as shown in Fig. 3.9 then a steady state orientation error, being $\Delta\theta^{ss}$, exists equal to

$$\begin{aligned}\Delta\theta^{\text{ss}} &\equiv \lim_{t\to\infty}\left({}^c_t\theta^d(t) - {}^c_t\theta^m(t)\right)\\ &= \lim_{s\to 0} s\left(\frac{s}{s+K_\theta}\cdot\frac{-{}_tv_y}{r.s^2}\right)\\ &= \frac{-{}_tv_y}{r.K_\theta},\end{aligned}\tag{3.5}$$

with $s/(s+K_\theta)$ the closed loop transfer function from input to orientation error and $-{}_tv_y/(r.s^2)$ the input signal.

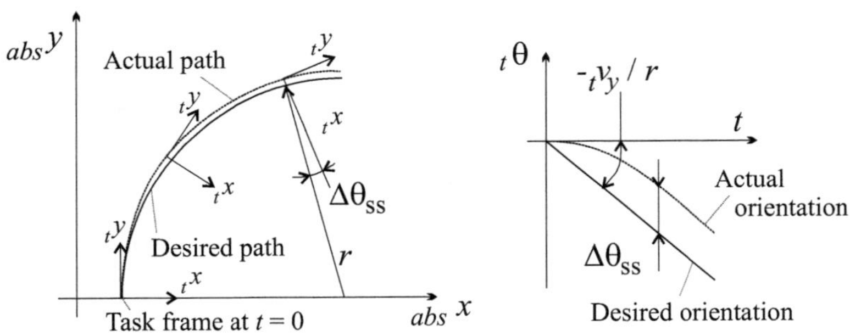

Fig. 3.9. 2D path (**left**) and time response (**right**) for the proportional control of the orientation of the task frame following an arc shaped contour with fixed (desired) tangent velocity.

The orientation control loop is independent of the position control loop, since moving the task frame in the ${}_tx$-direction does not change the angle ${}^c_t\theta$. In contrast, the position measurement (and control) depends on the orientation of the task frame, and thus on the orientation control loop.

Once the orientation control loop reaches an equilibrium, ${}^c_t\theta$ no longer changes and is equal to $\Delta\theta^{\text{ss}}$. Hence in steady state, the position control loop sees the contour at a constant angle $\Delta\theta^{\text{ss}}$. This is equivalent to the position control of a non-rotating task frame with a straight contour. The position tracking error[8] thus follows from (3.2) and (3.5):

$$\Delta x^{\text{ss}} = \frac{-{}_tv_y\tan(\Delta\theta^{\text{ss}})}{K_x} \approx \frac{-{}_tv_y\Delta\theta^{\text{ss}}}{K_x} = \frac{{}_tv_y^2}{rK_\theta K_x}.\tag{3.6}$$

The simulation results, shown in Fig. 3.10, confirm this equation. The actual path of the task frame in steady state follows from three simultaneous, fixed velocities, being ${}_tv_y$, ${}_tv_x^{ss}$ and ${}_t\omega^{ss}$:

$$ {}_tv_x^{ss} = K_x \Delta x^{ss} \approx \frac{{}_tv_y^2}{rK_\theta} \quad \text{and} \quad {}_t\omega^{ss} = K_\theta \Delta\theta^{ss} = \frac{-{}_tv_y}{r}. \tag{3.7} $$

Equation (3.7) is consistent with the results of (3.6) and (3.5).

Appendix A, Sect. A.6, describes the simulation scheme for the arc shaped contour following control with a rotating task frame and gives a detailed analysis of the actual task frame motion in steady state. The simulation results of which some are given in Fig. 3.11, correspond to the theoretical equations. The transient behaviour of the orientation tracking error is of first order. The position tracking error shows a second order behaviour.

If we add an additional integrator to the control loop of Fig. 3.8, the steady state orientation tracking error $\Delta\theta^{ss}$ will become zero, provided the stability of the control loop is maintained. However, a better solution to eliminate any steady state tracking error, not affecting the stability, is again the use of

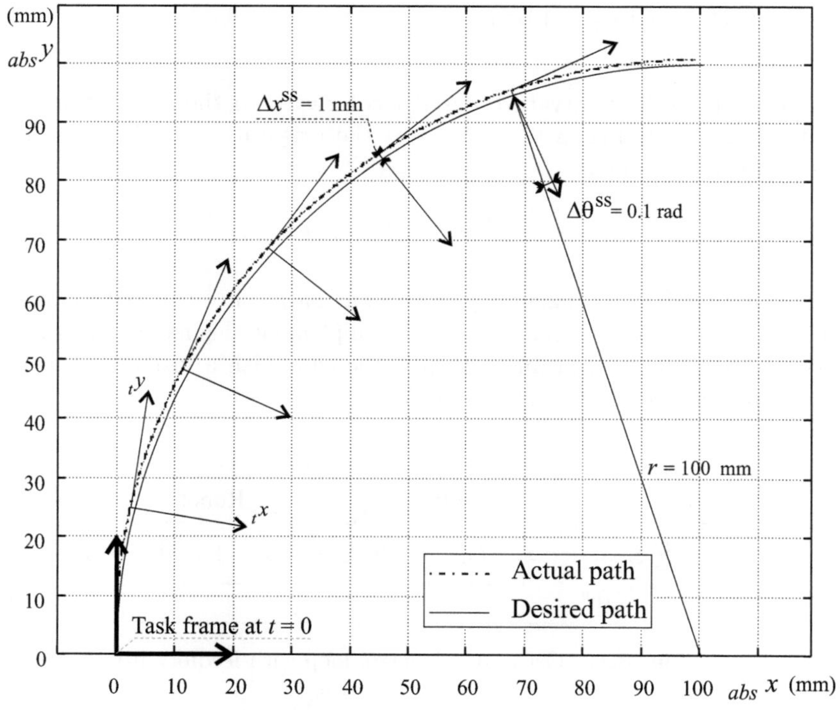

Fig. 3.10. 2D path of a rotating task frame, following an arc shaped contour with radius 100 mm, without feedforward control, with $K_x = 1/\text{s}$, $K_\theta = 1/\text{s}$ and ${}_tv_y = 10\,\text{mm/s}$

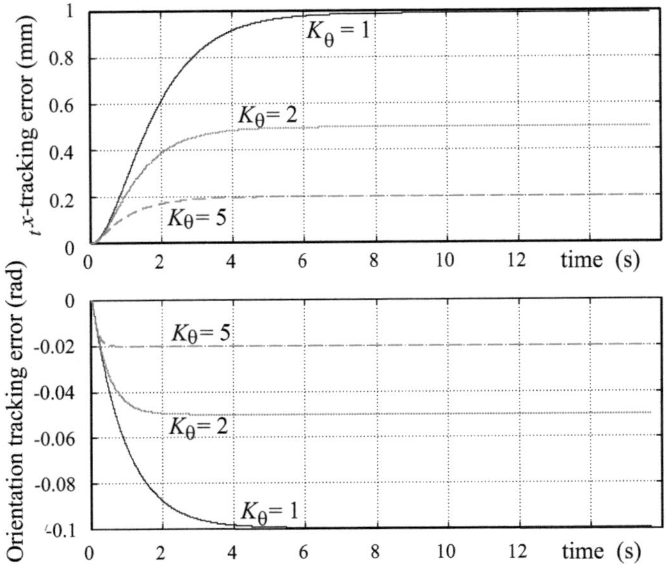

Fig. 3.11. Transient behaviour of position and orientation tracking errors in following an arc with radius 100 mm, without feedforward control, with $K_x = 1/\text{s}$, K_θ = variable (1/s) and $_tv_y = 10\,\text{mm/s}$

feedforward. Adding curvature based feedforward to the orientation control loop, as shown in Fig. 3.12, eliminates the orientation tracking error. The feedforward component

$$_t\omega^{\text{ff}} = -\frac{_tv_y}{r} = \kappa . _tv_y \tag{3.8}$$

keeps the task frame tangent to the circular path, once this path is reached. This too is confirmed by simulations. To implement this feedforward the curvature κ must be measurable, which is the case with a vision sensor, or the curvature must be known from a model.

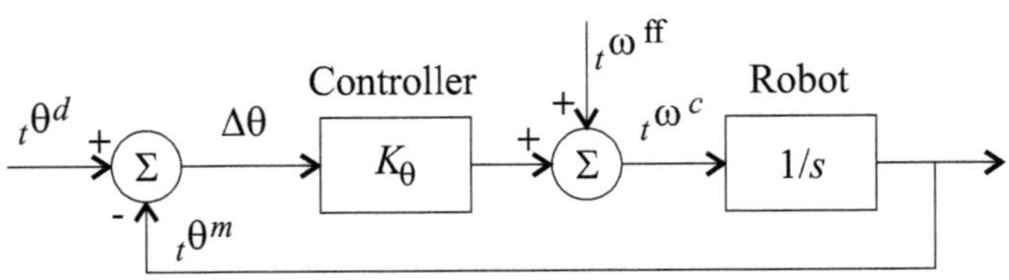

Fig. 3.12. Orientation control loop with feedforward

According to (3.6), a zero orientation tracking error also implies that the position tracking error Δx^{ss} becomes zero in steady state.

3.2.5 Conclusion

This section describes the steady state errors involved in planar contour following. Since the control loops (for position and orientation) contain one integrator, there are constant tracking errors for the non-rotating task frame following a straight contour and for the rotating task frame following an arc shaped contour.

The use of feedforward can eliminate the steady state tracking errors without affecting the control stability. For the position control loop the feedforward component is tangent based. For the orientation control loop the feedforward component is curvature based. Both orientation and position tracking errors become zero in steady state when following a contour with constant curvature by adding curvature based feedforward.

Although the static characteristics are derived only for the case of 2D planar contour following (3DOF), the conclusions are, in view of the one-dimensional control loop analysis, equally valid for 3D or 6DOF cases[10].

3.3 Dynamic Control Aspects

This section investigates the dynamic characteristics of the vision and force control loops. The theoretical control problem is simplified to the design of one-dimensional control loops.

In contrast to most robot controllers our robot accepts velocity set-points instead of position set-points. This makes a trajectory generation, as is well known in robotics, superfluous. In our approach, a perceived error generates a velocity command which moves the robot towards the target. This automatically results in a target following or contour following behaviour without the complexities of an explicit trajectory generator [18, 20, 21].

In addition, using velocity set-points as input lowers the order of the control loop: after all, when using velocity set-points as input, the robot itself forms a natural integrator, hereby eliminating steady state errors in the loop that encloses the robot. As shown in the previous sections, there is no need for an additional integrating action in the controller, if feedforward is used. In a position controlled robot, on the other hand, the robot itself is no longer an integrator. To eliminate steady state errors, the outer control loop needs to possess an integrator. The redundant levels of control add to system complexity and may have impact on the closed-loop performance [17, 20].

Vision and force control are clearly different. The bandwidth of the vision control loop is limited by the video frame rate of 25 Hz[11]. In theory, the

[10] This conclusion does not apply to the sensors, e.g. torsion (in 3D) is not measurable by a (mono-)vision system.

[11] With separate use of odd and even video frames, the frame rate becomes 50 Hz.

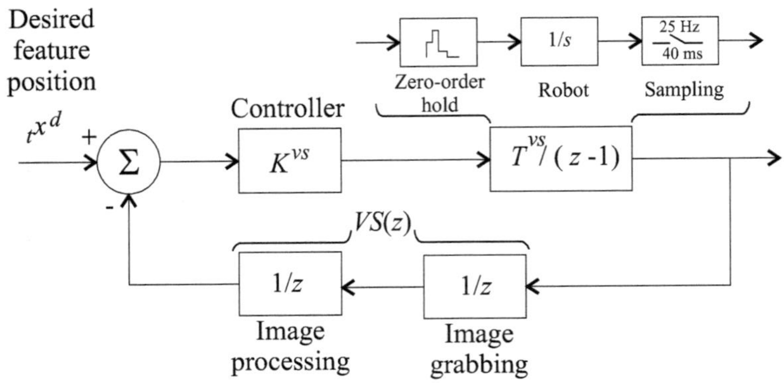

Fig. 3.13. One-dimensional vision control loop at 25 Hz ($T^{vs} = 40$ ms)

bandwidth of an analogue force sensor is much larger. In practice, our force control loop runs at a frequency of only 100 Hz. Unfortunately, the available computational power does not allow higher sample frequencies. Moreover, the force measurements are (very) noisy. This enforces the need of a filter which is implemented digitally at the same sample frequency of 100 Hz! The maximum possible cut-off frequency of the force measurement filter is hence 50 Hz. As Sect. 3.4 will show, the noise level involved in the experimental force measurements compels to the use of cut-off frequencies of 10 Hz or lower! For both vision and force control loops, the previous arguments mean that the dynamics of the inner joint control loop of the robot are negligible. Hence, as was the case in Sect. 3.2, for the dynamic analysis too, the complete robot may again mainly be modelled as an integrator.

3.3.1 Vision Control Loop

Figure 3.13 shows the simplified one-dimensional digital vision control loop. The sample frequency is equal to the video rate, being 25 Hz. This means that each 40 ms, which is the sampling period T^{vs}, the input for the robot is updated. The proportional control gain is K^{vs}. The robot (with joint controller) is modelled as an integrator ($R(s) = 1/s$). Using a zero order hold function on the input of the robot, the discrete equivalent of the robot ($R(z)$) equals

$$R(z) = T^{vs}/(z-1), \tag{3.9}$$

with z the Z-transform variable. The vision sensor $VS(z)$ is modelled by a (pure) delay. The total delay in the computation of the image features, consists of two components: the image grabbing delay on the one hand and the image processing delay on the other. Grabbing a full size image, seizes one time period T^{vs}. The processing delay varies from a few ms to maximum one time period T^{vs}. In order to be able to work with a fixed delay, the maximum total delay of $2 * T^{vs}$ is taken, (see also Fig. D.1 of Appendix D), hence

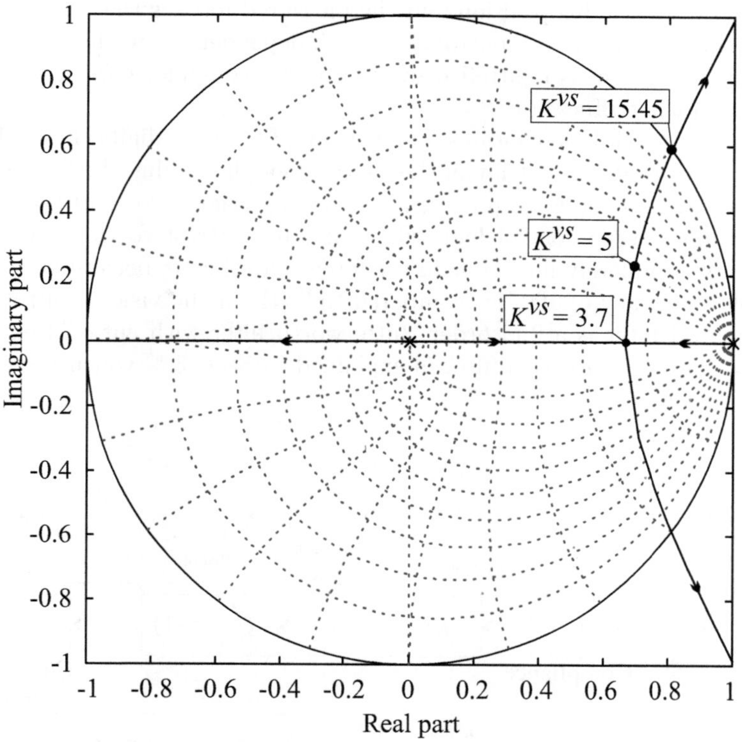

Fig. 3.14. Root locus for the digital vision control loop of Fig. 3.13

$$VS(z) = z^{-2}. \tag{3.10}$$

Figure 3.14 gives the root locus for the vision control loop. For a gain $K^{vs} = 1$ (1/s), the Gain Margin (GM) is 15.45. The closed loop response is critically damped for $K^{vs} = 3.7$. For a gain $K^{vs} = 5$ (GM $\simeq 3$), the closed loop step response shows 4% overshoot[12]. Taking the simplification of the model into account, we can conclude that the maximum (desirable) proportional control gain in a vision direction is 5/s. Good values range from 2 to 5.

3.3.2 Force Control Loop

Figure 3.15 shows the simplified one-dimensional digital force control loop. The force sample frequency is 100 Hz or, in other words, every $T^f = 10\,\text{ms}$ the input to the robot is updated. The proportional control gain is K^f. The robot $R(z)$ is (again) modelled as an integrator, giving ($R(z) = T^f/(z-1)$) as the zero order hold equivalent. The output of the robot is a position. The

[12] Defining the bandwidth of the control loop as the 3 dB attenuation frequency results in a closed loop bandwidth of 1 Hz for $K^{vs} = 3.7$ and 1.75 Hz for $K^{vs} = 5$.

relation between this position and the measured force is the contact stiffness k (N/mm). To make the control gain K^f independent from the contact stiffness, the force error is multiplied by the contact compliance k^{inv}, which is the inverse of the stiffness ($k^{inv} = k^{-1}$).

The feedback loop contains a unit delay (T^f) and a digital filter. The unit delay incorporates any delay in the control loop, including, the force measurement delay and the force processing. It mainly reflects some of the dynamics of the inner servo control loop which are, in contrast to the vision control loop, not completely negligible in the force control case, because of the higher sample frequency of 100 Hz (compared to 25 Hz in the vision control case)[13]. The digital filter is a 2nd order Butterworth filter with cut-off frequency at 2.5 Hz. The need for this filter is explained in Sect. 3.4, which discusses the force sensor.

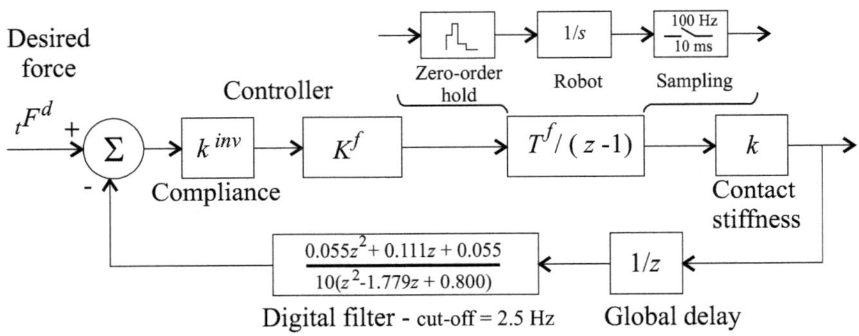

Fig. 3.15. One-dimensional force control loop at 100 Hz ($T^f = 10$ ms)

Figure 3.16 gives the root locus for the force control loop. For a gain $K^f = 1/s$, the Gain Margin (GM) is 16.9. The closed loop response approximates a critically damped behaviour[14] (no overshoot) for $K^f = 4$. For a gain $K^f = 5$, the closed loop step response shows 8% overshoot. In view of the simplified robot model, the maximum (desirable) proportional control gain is 4/s. Good values range from 2 to 4.

If the cut-off frequency of the digital filter increases (for example to 5 Hz or 10 Hz), the (theoretical) gain margin of the control loop will also increase and higher control gains are possible. But also the noise level will grow. This makes the practical control restless, especially if the control gains are set too high. Even more, the dynamics of the filter will (theoretically) no longer outshine the robot dynamics, such that these would need to be modelled more accurately.

[13] The introduction of a 'full' unit delay in the control loop may be a slight overestimation of the real delay and dynamics, but it will surely give a 'save' controller design.

[14] This corresponds to a closed loop bandwidth equal to 1.3 Hz.

The overall stability of the control loop will not change that much and the (newly) found control gains will come close to the ones given previously.

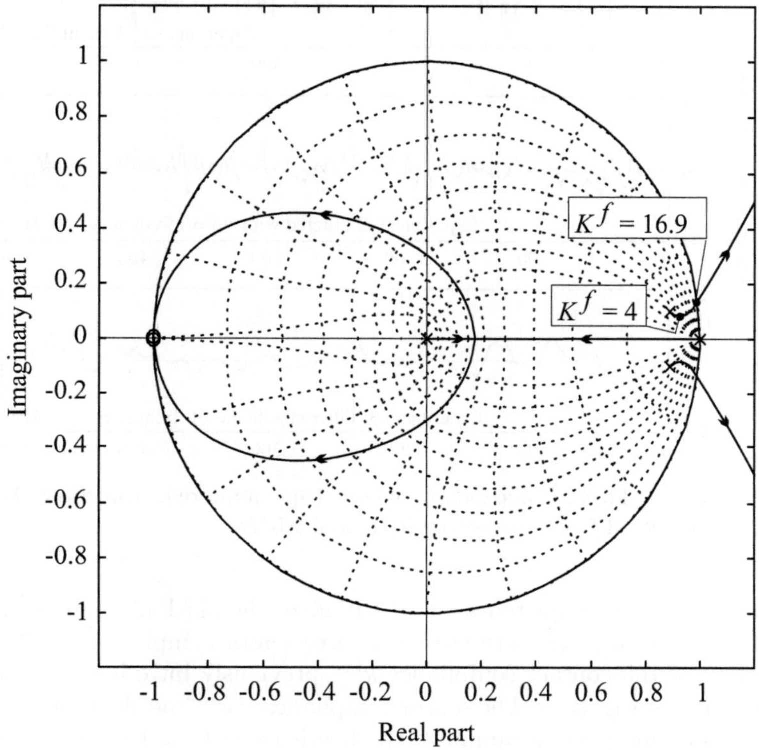

Fig. 3.16. Root locus for the digital force control loop of Fig. 3.15

3.4 Force Sensing

The used force sensor is a strain-gauge force sensor[15], which senses the acting forces by measuring the deformation of the sensor. The sensor output is a wrench: a 6×1 vector containing 3 forces and 3 moments. The measured forces, from environment onto the robot, in the force sensor frame are recalculated to their equivalents in the task frame by the force screw transformation ${}^{fs}_{t}S^{f}$ as explained in Sect. A.2.

A force sensor calibration determines the scale factors (between actual measured signals and corresponding forces), the mass of the tool and the center point of this mass. The latter two parameters are used to compensate gravity forces.

[15] Type: SCHUNK, Model: FT 450/45, Sensor range: force ± 450 N, torque ± 45 N m

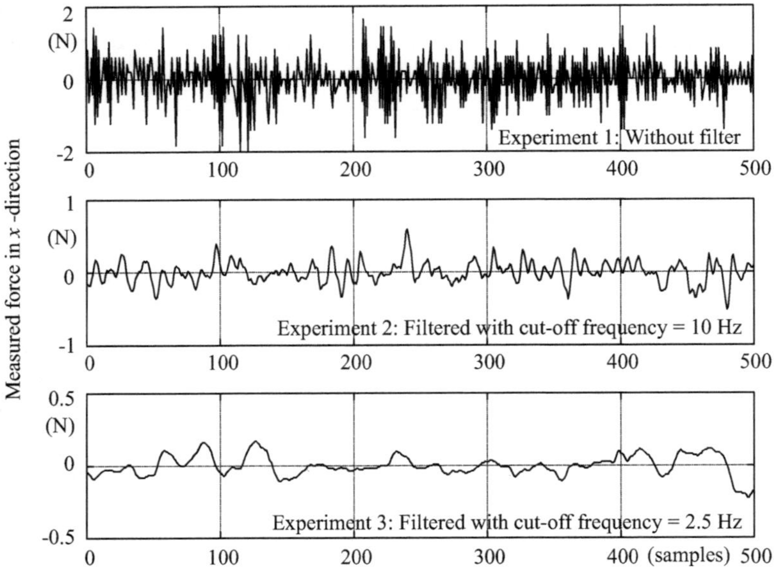

Fig. 3.17. Noise level of force measurement: (**top**) unfiltered, (**middle & bottom**) filtered with cut-off frequency set to 10 Hz and 2.5 Hz

For reasons of completeness, Fig. 3.19 gives the block diagram of the force sensor[16]. If strain gauges are used, the force sensor compliance ${}^{fs}k^{inv}$ is only a fraction of the contact compliance k^{inv} previously introduced in the force control loop of Fig. 3.15. The sensor compliance ${}^{fs}k^{inv}$ needs to be very small (stiff sensor) in order to minimize the loading effects of the sensor onto the system, as is well known from measurement theory.

Figure 3.17 shows the noise level involved in the force measurements. The top figure gives the unfiltered measurement data. The peak to peak noise level is about 3.5 N[17]. Peaks on the unfiltered forces up to 10 N have been measured (with the robot actuators on, but standing still).

Due to the high noise level a low-pass filter with cut-off frequency at maximum 10 Hz is necessary. Frequently, the cut-off frequency is even decreased to 5 Hz or 2.5 Hz. The digitally implemented low pass filter is a 2nd order Butterworth filter. Figure 3.17 (middle & bottom) shows the remaining noise

[16] Figure 3.19 is simplified in one aspect. In the transmission of the measured signal to the controller unit, the digital data is made analogue, connected to a standard analogue input of the controller board and then resampled. This makes it possible to use different (force) sensors in the same 'way' with the disadvantage, however, that the signal quality degrades.

[17] The high noise level is partly explained by the fact that the force sensor's measuring range (= 450 N) is rather large. For information, the sensor resolution is about 0.35 N.

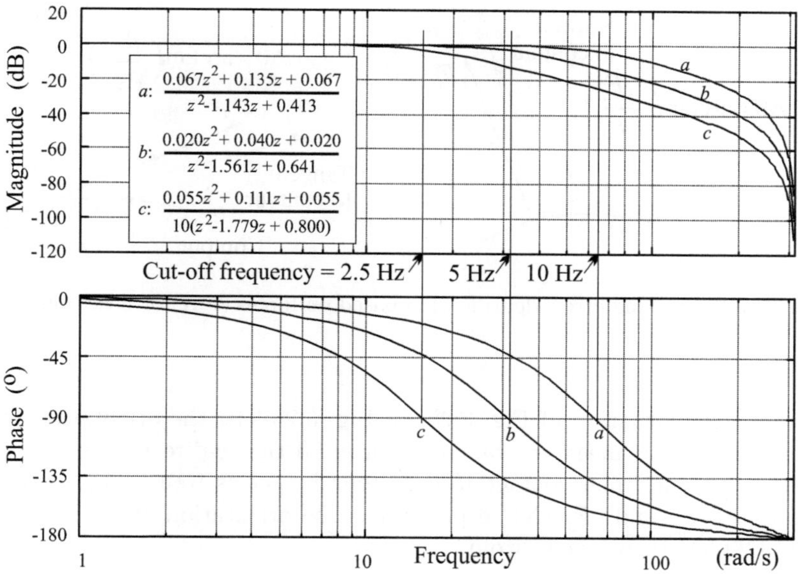

Fig. 3.18. Bode plots of 2nd order digital Butterworth filters with cut-off frequency at (**a**) 10 Hz, (**b**) 5 Hz or (**c**) 2.5 Hz at a sampling rate of 100 Hz

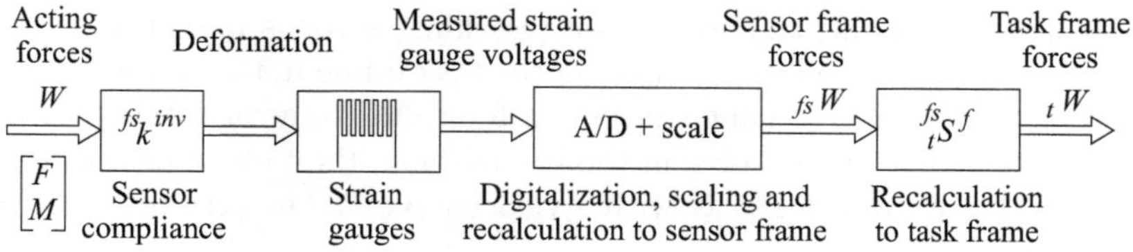

Fig. 3.19. Block diagram of force sensor

level in the force measurement using these filters. Figure 3.18 shows the corresponding Bode plots (and equations) of three Butterworth filters with cut-off frequencies at 10 Hz, 5 Hz and 2.5 Hz for a sample rate of 100 Hz.

Without a filter, the robot behaves restlessly and is not adequately controllable. On the other hand, the cut-off frequency is so low that the proportional control gains highly depend on the filter (see previous section)!!

3.5 Tracking Error Identification

This section discusses the identification of the tracking error in detail. The tracking error identification is either based on measured forces or based on (commanded) velocities. In the former case, we call the corresponding control action *tracking on forces*, in the latter case, *tracking on velocities*.

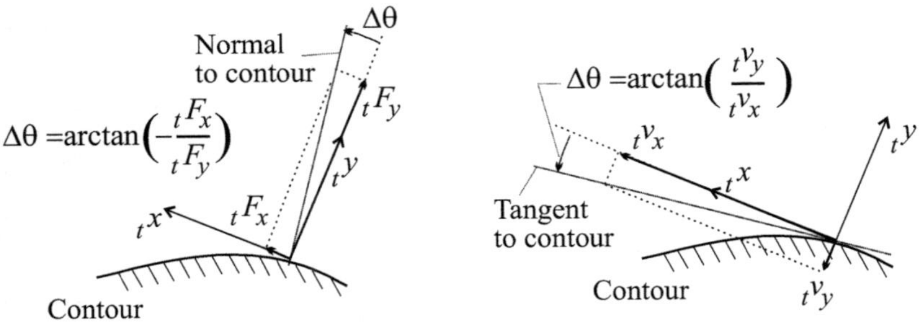

Fig. 3.20. Tracking error identification based on forces (**left**) or based on velocities (**right**)

Take the situation of Fig. 3.20. In Fig. 3.20-left, the orientation error of the task frame is measured based on forces. If the ${}_ty$-direction is not tangent to the contour, then contact forces occur in ${}_tx$- as well as in ${}_ty$-direction. They are ${}_tF_x$ and ${}_tF_y$ (N) respectively. The orientation error $\Delta\theta$ shown in Fig. 3.20-left then follows from

$$\Delta\theta = \arctan\left(-\frac{{}_tF_x}{{}_tF_y}\right) \tag{3.11}$$

In Fig. 3.20-right, the orientation (tracking) error is measured by examining the task frame velocities. Moving the task frame in the ${}_tx$-direction, with a velocity ${}_tv_x$, causes a change in contact force due to the orientation error $\Delta\theta$. To maintain the contact force in the ${}_ty$-direction, the task frame has to move towards the contour, let's say with a velocity ${}_tv_y$[18]. The orientation error $\Delta\theta$ then follows from

$$\Delta\theta = \arctan\left(\frac{{}_tv_y}{{}_tv_x}\right) \tag{3.12}$$

If the ${}_tx$-direction lies tangent to the contour, the orientation error is equal to zero.

Identification based on velocities is badly conditioned when moving slowly (${}_tv_x \approx 0$ in (3.12)) and is disturbed by the deformation of the tool. On the other hand, friction forces, which are difficult to model, disturb the identification of the tracking error based on forces. Both identification methods are noise sensitive. In the velocity based identification, however, we use the commanded ${}_tx$-velocity instead of the measured one, hereby making the identification less noise sensitive. Hence, if the tangent velocity is not too small, the velocity based identification of the tracking error will give the best results.

[18] The velocity ${}_tv_y$ in the force controlled direction equals $K_y^f k_y^{inv}({}_tF_y^m - {}_tF_y^d)$ and thus depends in se on the measured forces.

3.6 Relative Area Contour Detection

The following describes the used low level image processing techniques: the relative area contour detection (this section) and the ISEF edge detection (Sect. 3.7).

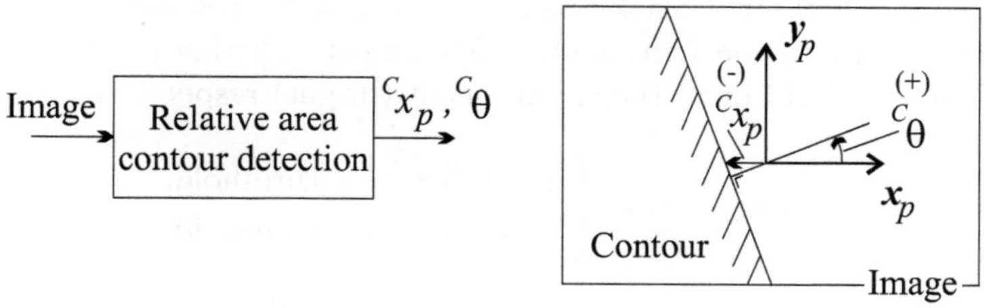

Fig. 3.21. Block diagram of relative area contour detection (**left**) and definition of contour feature parameters (**right**)

The relative area (or pixel-weighting) method is a *direct method.* It measures the image features of interest directly from the thresholded image without any intermediate steps. These image features are the distance to the contour from the center of the image $^{C}x_p$ (pix) and the angle of the contour $^{C}\theta$ (rad) as shown in Fig. 3.21-right. Figure 3.21-left shows the single block diagram of this method, emphasizing its direct characteristics[19].

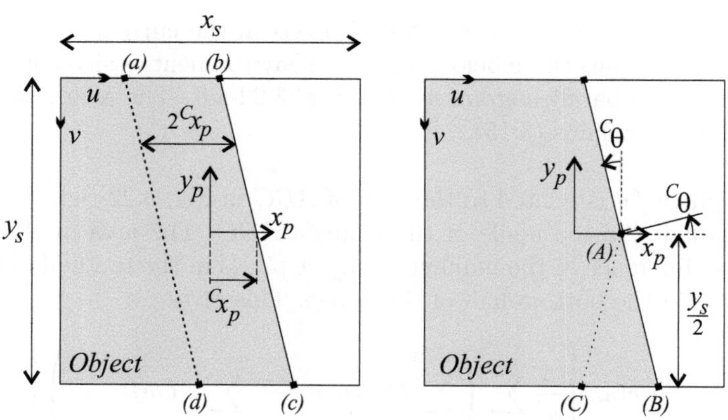

Fig. 3.22. The position $^{C}x_p$ of the straight contour is proportional to the area *abcd* (**left**); the orientation of the contour $^{C}\theta$ is related to the area *ABC* (**right**).

[19] This in contrast to the ISEF edge detection method, described in Sect. 3.7 for which the found edge positions are only an intermediate step in the measurement of the contour.

Figure 3.22 illustrates the principle. Under the assumption of a straight contour which enters the image through the bottom-side and leave the image through the top-side, ${}^{C}x_p$ and ${}^{C}\theta$ follow from the weighting of 'positive' and 'negative' object or environment pixels[20]. The object is assumed to be dark, the environment bright[21].

Let $I(u,v)$ be the brightness (also intensity or grey-level value) of pixel (u,v), with $I = 0$ for a black pixel and $I = 255$ for a white one. Define the functions $Obj(u,v)$ and $Env(u,v)$, which indicate whether pixel (u,v) belongs to the object (dark) or to the environment (bright) respectively, as

$$Obj(u,v) = \begin{cases} 1 & \text{if } I(u,v) < \text{threshold}, \\ 0 & \text{if } I(u,v) \geq \text{threshold}, \end{cases} \tag{3.13}$$

and

$$Env(u,v) = 1 - Obj(u,v) = \begin{cases} 0 & \text{if } I(u,v) < \text{threshold}, \\ 1 & \text{if } I(u,v) \geq \text{threshold}. \end{cases} \tag{3.14}$$

The threshold value is chosen in such a way that it clearly separates object and environment, e.g. equal to 120 (medium grey).

Then, the distance ${}^{C}x_p$, for an x_s by y_s image, is given by:

$${}^{C}x_p = \frac{1}{2y_s}\left[\sum_{u,v=1,1}^{x_s,y_s} Obj(u,v) - \sum_{u,v=1,1}^{x_s,y_s} Env(u,v)\right] \tag{3.15}$$

Equation (3.15) counts the number of pixels in the current image belonging to the object minus those belonging to the environment, which corresponds to the area of the parallellogram *abcd* in Fig. 3.22-left. The area of *abcd* equals $2{}^{C}x_p y_s$. This explains (3.15).

The angle ${}^{C}\theta$ is related to the area of ABC in Fig. 3.22-right. ${}^{C}\theta$ is in fact equal to half the top angle of the triangle ABC. The area of ABC follows from the difference in the number of object pixels in the top half of the image and those in the bottom half of the image. This gives

$${}^{C}\theta = -\arctan\left[\frac{4}{y_s^2}\sum_{u=1}^{x_s}\left(\sum_{v=1}^{y_s/2} Obj(u,v) - \sum_{v=\frac{y_s}{2}+1}^{y_s} Obj(u,v)\right)\right]. \tag{3.16}$$

The main *advantage* of the relative area method is its robustness. No previous filtering of the image is needed because the summation over all object pixels will average out the effect of faulty pixels. It is also a very easy to implement and quite fast method.

[20] 'Pixel' originated as an acronym for PICture'S ELement.

[21] Mark that such a condition can always be achieved by the use of back-lighting.

There are however important *restrictions*. First, this method needs a threshold value, to separate the pixels belonging to the object from those belonging to the environment. This is a major drawback when used under changing lighting conditions.

Second, the accuracy of the position and particularly the angle measurement is limited due to the limited size of the image and the inevitable pixel quantization. Figure 3.23 gives an example. The left part of Fig. 3.23 shows the (simulated) measured angle versus the actual angle of the computer generated contour. It further shows the resulting error for angles going from 0° to 25°. The image size is 64 by 64 pixels. The right part of Fig. 3.23 gives the theoretical minimum (positive) measurable angle according to (3.16) as a function of the image height y_s. The smaller the image, the less accurate the orientation measurement will be.

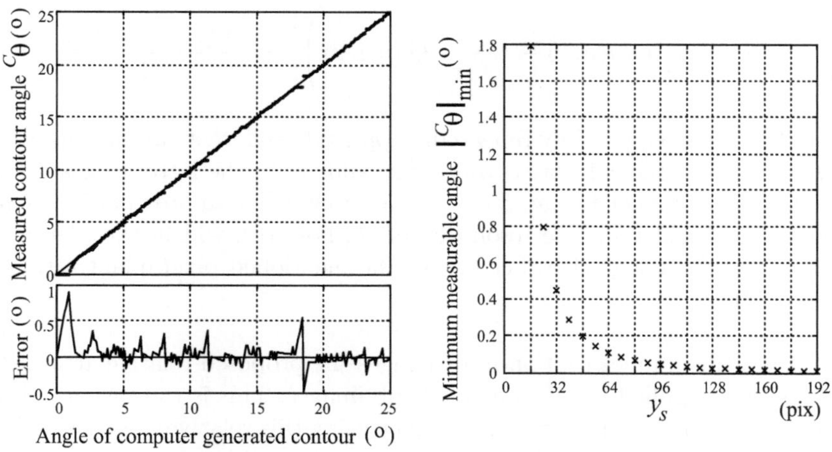

Fig. 3.23. Examples of the accuracy of the relative area contour measurement: (**left**) measured contour angle and error versus angle of 64 by 64 computer generated image; (**right**) absolute value of the minimum measurable angle as a function of the image height y_s

Third, as previously mentioned, (3.15) and (3.16) are based on the assumption that the contour enters and leaves the image through top- and bottom-sides[22]. This can be checked by looking at the four corners of the image. If the assumption is not true, the problem is solved by taking a sub-image for which it is true.

[22] Although (3.15) and (3.16) may not be valid for contours with large angles, the incorrect measurement will not affect the correctness of the positioning. The generated signal (with correct sign) will still result in a movement of the robot towards the desired image, containing a vertical contour, for which the equations are valid again.

Fourth, (3.15) and (3.16) are valid for vertical contours with the object lying left in the image. Equivalent equations for horizontal contours and/or the object to the right can be deduced.

Finally, the calculated parameters refer to a line. They are therefore only valid for a straight contour, resulting in a less accurate measurement (and hence positioning) with curved contours. An image of a straight contour, however, is point symmetric w.r.t. any frame centered at the contour. This property can be checked with following relation (see Sect. A.4 for the full deduction of this equation): An image with vertical contour is point symmetric if

$$\begin{aligned}&\sum_{v=1}^{y_s}\left(\sum_{u=1}^{\mathrm{floor}(\frac{x_s}{2}+{}^C x_p)} Obj(u,v) - \sum_{\mathrm{floor}(\frac{x_s}{2}+{}^C x_p)+1}^{x_s} Obj(u,v)\right) \\ &- \left|\sum_{u=1}^{x_s}\left(\sum_{v=1}^{y_s/2} Obj(u,v) - \sum_{v=\frac{y_s}{2}+1}^{y_s} Obj(u,v)\right)\right| \simeq -\frac{(x_s + 2\,{}^C x_p).y_s}{2}\end{aligned} \tag{3.17}$$

This equation, possibly repeatedly applied to different sub-images, will validate or reject the calculated values ${}^C x_p$ and ${}^C\theta$. In the latter case a different method or a smaller image, for which a straight line approximation is valid, has to be considered. The resolution of ${}^C\theta$, however, will further deter when decreasing the image sizes because of the mentioned pixel quantization (see Fig. 3.23).

All of this implies that on most images the proposed relative area method can only be applied locally, on small sub-images. This is the case for chapter 5 which presents a successful experiment using this relative area method. The aim of this experiment is the alignment at a given distance of an uncalibrated eye-in-hand camera with an arbitrarily placed rectangle for the full 6 degrees of freedom.

3.7 ISEF Edge Detection

This section briefly discusses the used ISEF edge detector. As shown in Fig. 3.24, the ISEF edge detection is the first step in obtaining the necessary control data. The input for the ISEF edge detection is the raw image, the output is a set of contour edge points $[X_p, Y_p]$, which in their turn are the basis for the contour modelling and the control data extraction, described in the next section. The number of contour points is chosen to be sufficient (e.g. 7 to 11 points) for a good subsequent fit. Since the image features are computed in several steps, we call this approach an *indirect method*, in contrast to the (direct) relative area method of the previous section.

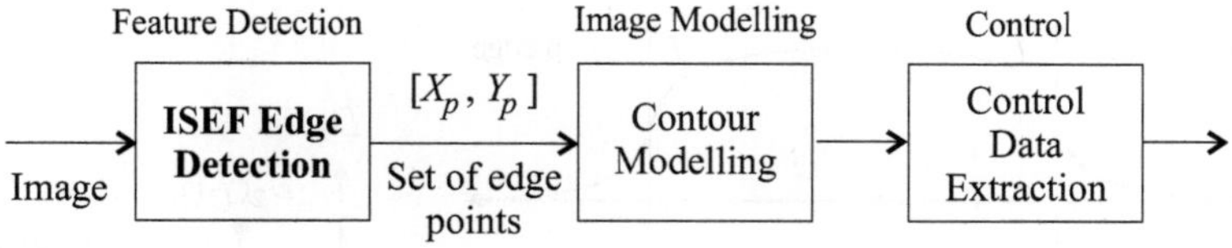

Fig. 3.24. Block diagram of control data computation: the first step is the ISEF edge detection

The ISEF edge detector, proposed by Shen and Castan [73], consists of an Infinite Symmetric Exponential (smoothing) Filter (ISEF - see Fig. 3.25), which efficiently suppresses noise, followed by a differential element to detect the changes. It computes the first and second derivatives of the image brightness (or intensity) function (I). The location of a step edge corresponds to an extremum in the intensity gradient or to a zero crossing of the second derivative. The recursive realization of the ISEF edge detector results in a line convolution operator which works on a row or column of grey values ($I(j)$) according to the following equations:

$$\begin{aligned} y_{rd}(j) &= (1 - b_{isef})I(j) + b_{isef}\, y_{rd}(j+1), \quad j = l-1 \dots 1 \\ y_{ld}(j) &= (1 - b_{isef})I(j) + b_{isef}\, y_{ld}(j-1), \quad j = 2 \dots l \\ D_1(j) &= y_{rd}(j+1) - y_{ld}(j-1), \quad j = 3 \dots l-2 \\ D_2(j) &= y_{rd}(j+1) + y_{ld}(j-1) - 2I(j), \quad j = 3 \dots l-2 \end{aligned} \tag{3.18}$$

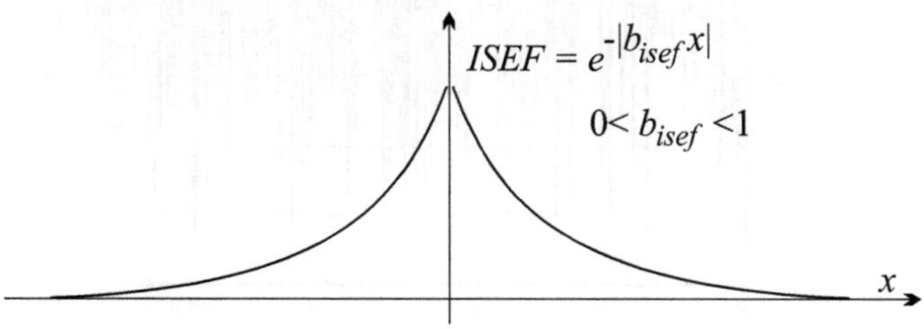

Fig. 3.25. Infinite symmetric exponential filter (ISEF)

y_{ld} and y_{rd} are left and right convolutions respectively; D_1 is the first derivative and D_2 is the second derivative; j is the pixel index on a row (or column); l is the length of the pixel row (or column); $I(j)$ is the brightness of pixel j and b_{isef} is the exponential constant, which shapes the exponential filter, shown in Fig. 3.25. Changing b_{isef} allows a trade off between noise suppression and edge location accuracy. Because the algorithm only uses line convolutions on a limited number of pixel rows, the real-time processing restrictions are easily met.

Figure 3.26-left gives an example of the edge detection. The (single) edge is pinpointed by the zero crossing of the second derivative at the extremum

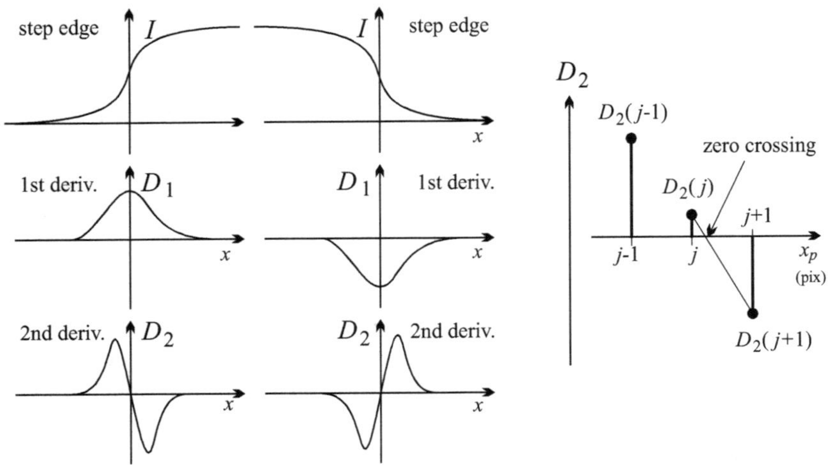

Fig. 3.26. Edge localization by the zero crossing of the second derivative: (**left**) brightness function, first and second derivatives; (**right**) linear interpolation to pinpoint the exact zero crossing (with sub-pixel accuracy)

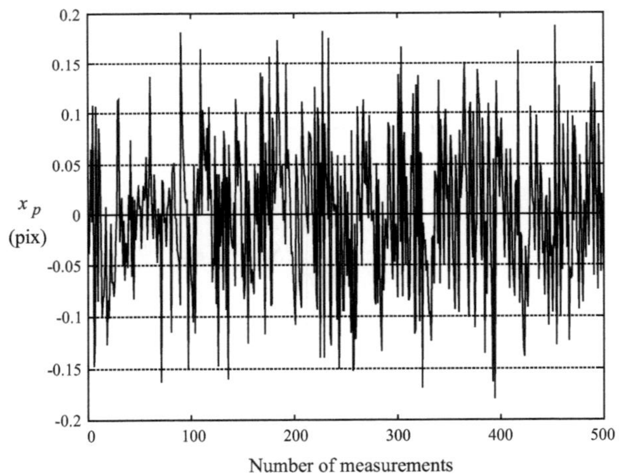

Fig. 3.27. Noise level of ISEF edge detector

of the first derivative of the image brightness function. If, due to the pixel quantization, the second derivative has no exact zero value, then the zero crossing follows from a linear interpolation as shown in Fig. 3.26-right. This results in sub-pixel accuracy for the edge localization.

Figure 3.27 shows the measured noise level for the implemented ISEF edge detector: A fixed image (with still camera) is grabbed and processed 500 times in a row; for each image, the ISEF detector measures the shown x-position of the edge on a fixed image row; the noise level is in the order of ± 0.15 pixels.

In contrast to the relative area algorithm, the ISEF edge detector is very sensitive to interlacing which occurs when the camera moves very fast (in a direction normal to the contour)[23]. As Fig. 3.28 exemplifies, such (undesirable) interlacing effects inevitably have impact on the accuracy of the contour measurement. Therefore, only odd image lines are used (or scanned). This avoids interlacing effects and hence improves the image processing.

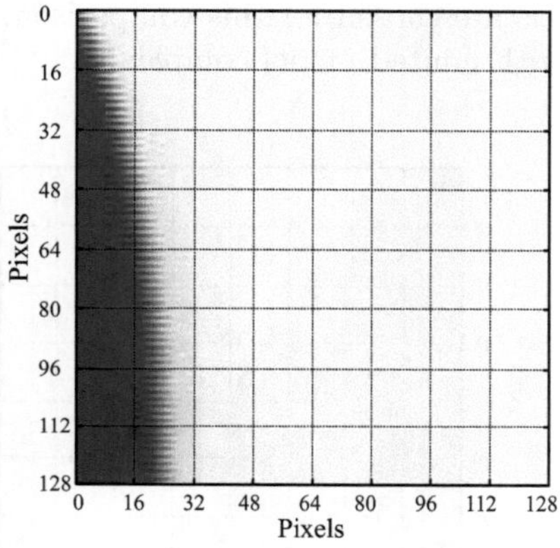

Fig. 3.28. Interlacing effect due to fast (sidewards) moving camera

3.8 Contour Modelling and Control Data Extraction

The previous section discusses the detection of edge points on a contour. This section investigates the contour modelling and control data extraction steps, shown in Fig. 3.29. The model is represented by the parameters of a function, which is fitted through the set of edge points. The searched control data, such as contour pose and curvature, are derived from the analytical representation of the contour.

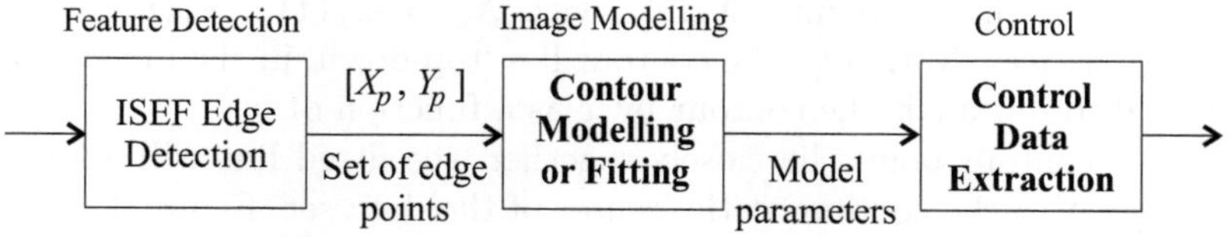

Fig. 3.29. Block diagram of contour modelling and control data extraction steps

Appendix B presents several contour models and evaluates them. The evaluation criteria (similar to those of Sect. 2.2) are:

[23] unless a progressive scan camera is used

1. the accuracy and robustness of the fitted function w.r.t. the real contour,
2. the suitability of the model as a basis for the control data computation and
3. the computational burden in obtaining the model and control data.

The following describes the tangent model and summarizes the conclusions of Appendix B by comparing the tangent model to two other contour models, being a parameterized third order polynomial and parameterized cubic splines. The latter two models are chosen for this comparison, because they give the best positional fits with limited model complexity.

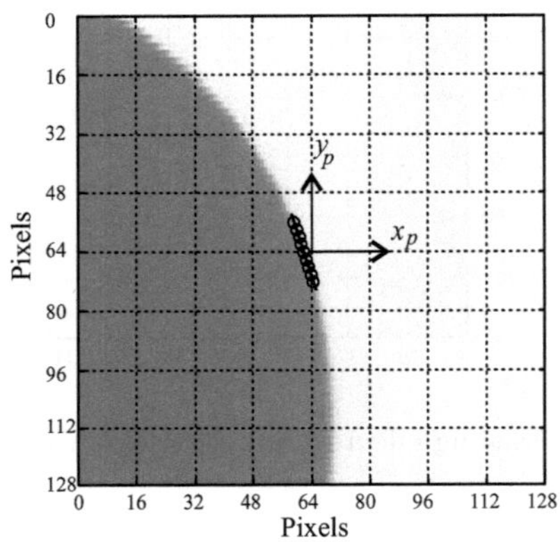

Fig. 3.30. Tangent to contour

3.8.1 The Tangent Model

The tangent contour model consists of (a set of) lines of the form

$$x_p = ay_p + b. \tag{3.19}$$

The model parameters a and b follow from a least squares fit of a line through the set of n contour points $[X_p, Y_p]$ with $X_p = [x_p(1) \ldots x_p(n)]'$ and $Y_p = [y_p(1) \ldots y_p(n)]'$. Normally, the contour lies 'top-down' in the image. Hence it is logical to represent the contour by x as a function of y.

If the contour points lie closely together, the fitted line will approximate the tangent to the contour at the center of the data set. Figure 3.30 gives an example.

The main advantage of this model is its simplicity. The contour pose[24] (position and orientation) for one single point on the contour is directly given by the model. A line, however, does not give any information about the contour curvature, at least not by itself. If, on the other hand, the tangents to

[24] similar to the output of the relative area method shown in Fig. 3.21

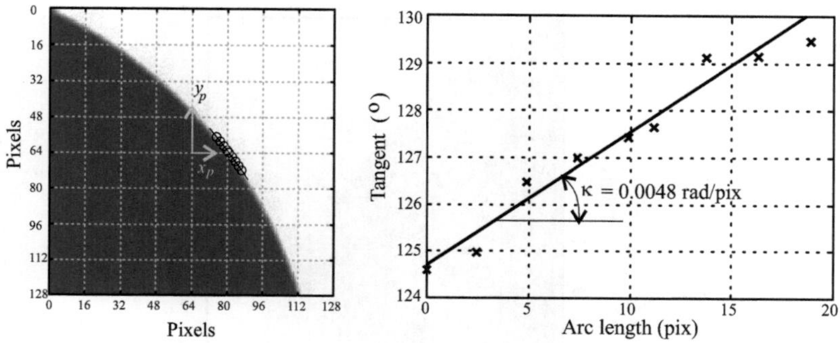

Fig. 3.31. Tangent to contour in nine points (**left**); Corresponding least squares solution for the curvature computation (**right**)

the contour are known for successive contour points, the curvature κ can be computed as the change in orientation ${}^{c}\theta$ as a function of the arc length s:

$$\kappa = \frac{d\,{}^{c}\theta}{ds}. \tag{3.20}$$

In practice, an approximation for κ follows from the least squares solution of a set of m first order equations

$${}^{c}\theta(i) = \kappa\, s(i) + cte, \quad i = 1 \ldots m. \tag{3.21}$$

The m (e.g. 9) pairs $({}^{c}\theta(i), s(i))$ lie symmetrically around the position of interest. Figure 3.31 gives an example. An experimental validation of the curvature computation according to (3.21) is given in Appendix C.

3.8.2 Comparison

Figure 3.32 compares the accuracy and suitability of three contour models. The input image, given at the top of the figure, is a computer generated circle with a radius of 209 pixels. A circular test contour is chosen, since it has a known constant curvature, in this case equal to $(1/209 =)$ $4{,}78.10^{-3}$ (1/pix). The contour is approximated by (1) tangent fits, (2) a parameterized 3rd order polynomial fit and (3) parameterized (smoothed) cubic splines. A full description of these contour models, among others, is given in Appendix B.

All three models are fairly robust and give a good positional approximation of the contour. The contour plots (top figure) are hardly distinguishable. For the computed contour tangent (middle figure), both the tangent model and the splines give good results. The polynomial model gives a less accurate tangent profile. But the most important difference emerges in the computed curvature profile given in Fig. 3.32-bottom.
Both the polynomial fit and the splines show unacceptable deviations in the computed curvature from the real one. Only the tangent model gives satisfactory results for the curvature, even in spite of the fact that the curvature is

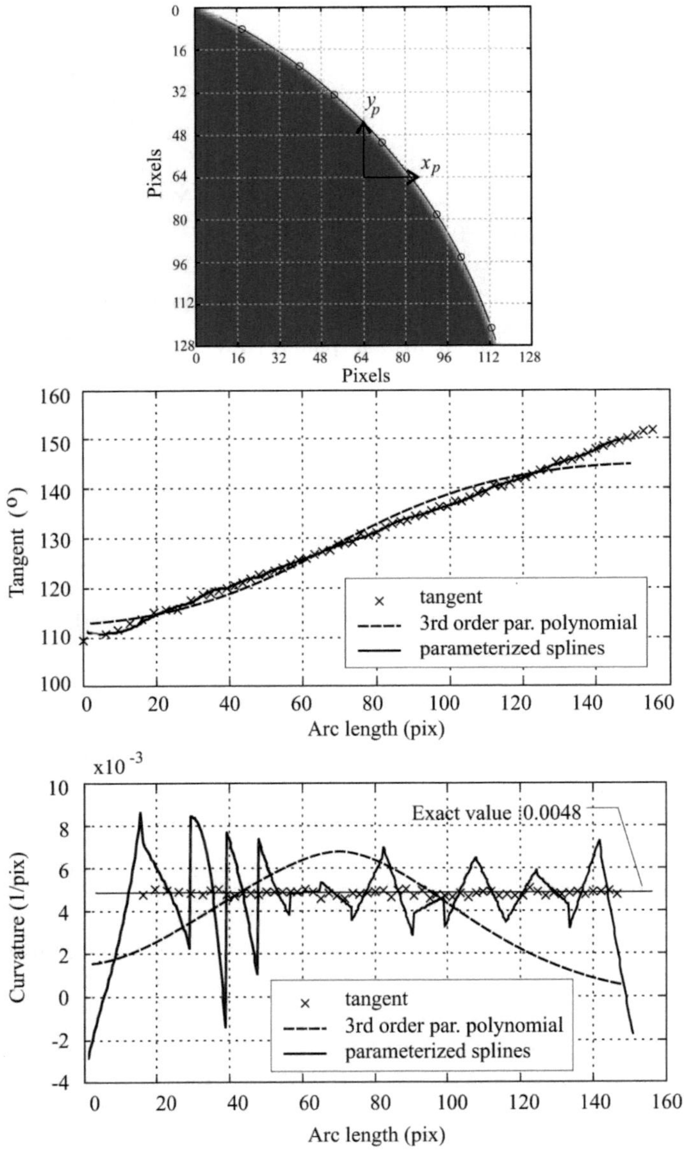

Fig. 3.32. Computer generated circle with 209 (pix) radius (**top**); Computed tangent (**middle**) and curvature (**bottom**) versus arc length for the arc shaped contour of the top image based on three different models: tangent fits, a parameterized 3rd order polynomial fit in 8 contour points and parameterized (smoothed) cubic splines with 17 contour knots

related to the second derivative of the position and thus difficult to compute accurately. The proposed (double) least squares solution clearly suppresses the explosion of noise which is inevitably related to a derivative operation.

The computational effort for the three models, leaving the parameterized character aside, is of the same order of magnitude: The tangent model uses two least squares fits, each in two unknowns with about 9 equations; a non parameterized 3rd order polynomial model uses one least squares fit in four unknowns with up to 8 (or more) equations and straightforward formulas for tangent and curvature computation; and the cubic splines are based on a set of 17 segments each using only straightforward formulas both for the three unknowns as well as for the tangent and curvature computation.

For a good fit on an arbitrary contour, however, both the polynomial and the splines need to be parameterized (see Appendix B), hereby doubling the computational effort, since both x and y are now expressed as a function of the parameter w. Even more, parameterization introduces a new problem: the determination of the parameter value for the contour position at which the curvature needs to be computed. For example, in the determination of the curvature at the contour position $y = 0$, we first need to compute the value of the parameter w for which $y = 0$ by searching the roots of the function $y = \eta(w)$. This again increases the computational effort. All of this implies that also for the computational effort, the tangent model is preferred over the other proposed models.

Table 3.2. Comparison of the characteristics for three contour models.

Accuracy	Tangent model	Third order parameterized polynomial	Parameterized smoothed cubic splines
$\hookrightarrow$ Position	Very good	Good	Very good
$\hookrightarrow$ Tangent	Very good	Poor	good
$\hookrightarrow$ Curvature	Good	Bad (unacceptable deviations)	
Order of points	Important for curvature	Not important	Very important -critical
Pose computation	Least squares solution	Twice least squares solution	Simple equations for each segment
Curvature computation	Least squares solution	Simple equations but extra parameter search	

3.8.3 Conclusion

Of all the investigated contour models, going from a simple tangent model over full second order circular or elliptic models to polynomials and splines possibly parameterized[25], the tangent model gives the most accurate approximation of the contour position, orientation and curvature at a limited computational effort, thus permitting a 'real time' implementation. Table 3.2 summarizes the characteristics of the three treated contour models.

3.9 Camera Model and Calibration

Finally, the camera calibration, explained in this section bridges the remaining gaps between image and real world coordinates on the one hand and between camera frame and task frame signals on the other.

3.9.1 Used Camera Model

The used camera model is the perspective view, pin-hole model shown in Fig. 3.33. Following equations describe the mapping of a point to its corresponding camera frame position by the pin-hole model:

$$\begin{cases} {}^{ap}_{cam}x &= -{}^{ap}_{cam}z \; {}^{ap}x_p \; \mu_p/f \\ {}^{ap}_{cam}y &= -{}^{ap}_{cam}z \; {}^{ap}y_p \; \mu_p/f \end{cases} \tag{3.22}$$

with f the focal length (mm), μ_p the effective pixel dimension (mm/pix) of the square pixels, $({}^{ap}x_p\,, {}^{ap}y_p)$ the image plane coordinates of an arbitrary point ap (pix) and ${}^{ap}_{cam}P = ({}^{ap}_{cam}x,\ {}^{ap}_{cam}y,\ {}^{ap}_{cam}z)'$ the camera frame coordinates of point ap (mm).

The given model has one fixed intrinsic camera constant (μ_p) and nine parameters: *three internal* (also called intrinsic or interior) parameters and *six external* (also called extrinsic or exterior) *parameters.*

The internal parameters[26] are the focal length f and the center coordinates of the image (C_x, C_y). The center of the image is expressed in the (u, v) coordinates: the image plain coordinates of a point in pixels relative to the top left corner of the image. (C_x, C_y) is the piercing point of the camera frame z-axis with the image plane, and thus the origin of the image coordinate frame $(x_p\,, y_p)$ (see Fig. 3.33).

The external parameters determine the position of the camera frame in a reference frame, which is in our case the end effector frame. They are (T_x, T_y, T_z) the translational components in mm and $(\alpha_x, \alpha_y, \alpha_z)$ the rotation angles in radians for the transformation between reference and camera frames.

---

25 See Appendix B for a full discussion of these models.

26 A complete model also includes the scaling factor s_x and the lens distortion κ_1 as internal parameters. See paragraph *Tsai's model and parameters.*

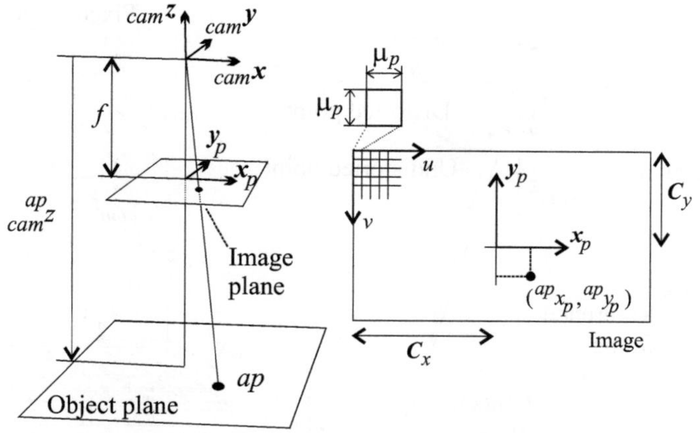

Fig. 3.33. Used Camera model with definition of camera frame, camera parameters and image plane coordinates

3.9.2 Calibration

Full calibration (for internal and external parameters) of the given camera model is essential in two cases:

- in position based visual servoing and
- in endpoint open-loop combined mounting of vision and force sensors.

In the former case, the calculation of the Cartesian position of an image feature uses by definition a known and calibrated camera model. In the latter case, calibration is essential to link the vision measurements to the contact (position) between force probe and object. In this work, Tsai's calibration technique [52, 80, 81] is adopted, however, with a substantial change in the setup.

3.9.3 Tsai's Model and Parameters

Tsai's camera model is a pin-hole model of 3D-2D perspective projection with first order radial lens distortion as shown in Fig. 3.34-left. The reference frame is an absolute fixed coordinate frame. The model has two additional internal parameters:

κ_1 : the radial lens distortion coefficient (mm^{-2}) and
s_x : a scale factor to account for any uncertainty in the framegrabber's resampling of the horizontal scanline.

The radial lens distortion coefficient κ_1 and the scale factor s_x, however, are of more importance in 3D vision measurements with high accuracy than in visual servoing. They are not used in our approach ($s_x = 1$ and κ_1 is neglected).

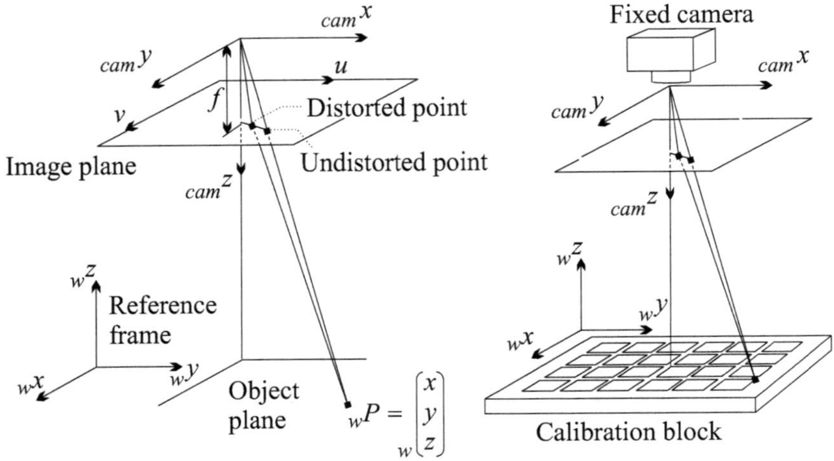

Fig. 3.34. **(left)** Tsai's pin-hole camera model: a geometric model of 3D-2D perspective projection with radial lens distortion; **(right)** Tsai's calibration setup: the corners of each square with exactly known world coordinates make up the calibration points

The first stage of Tsai's calibration solves a linear set of equations, using a least squares method, in five unknowns, $\alpha_x, \alpha_y, \alpha_z, T_x$ and T_y being the 3D camera position up to a variable distance in the z-direction. The second phase determines the remaining six unknowns in a recursive way. This is necessary due to the remaining non-linear equations. The division of the calibration problem in a linear and a non-linear part, is one of the main contributions of Tsai's calibration technique resulting in enhanced accuracy, speed and versatility.

3.9.4 Tsai's Setup

The calibration data for the model consist of the 3D (x, y, z) world coordinates of a feature point and the corresponding coordinates (u, v) of the feature point in the image. The quality of the calibration depends among other things on the accuracy at which the world coordinates of the calibration points are known. Tsai uses a calibration block with a pattern of exactly known squares and a fixed camera (see Fig. 3.34-right). The corners of each square, measured with sub-pixel accuracy, make up the calibration points. The positioning of the calibration block has to be performed very accurately. In a normal situation, this is a tedious and unpleasant operation. Hence, many researchers tend to avoid calibration.

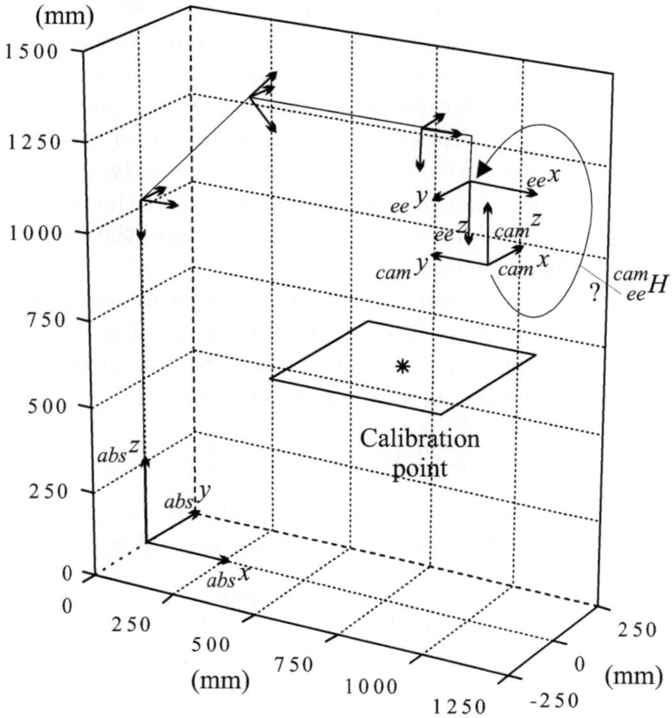

Fig. 3.35. View of the calibration setup (for one position) using only one calibration point with a fixed absolute position and a end effector mounted camera

3.9.5 Used Setup

Our setup highly differs from Tsai's setup. In contrast to Tsai, not the absolute pose of a fixed camera but the relative pose of an end effector mounted camera w.r.t. the end effector needs to be calibrated. Furthermore, our calibration procedure utilizes only one calibration point, thus avoiding the accurate positioning, mainly in orientation, of a calibration pattern. Figure 3.35 gives an overview of the used setup, defining absolute, end effector and camera frames. These frames are indicated by the preceding subscript *abs*, *ee* and *cam* respectively. The z-direction of the camera frame coincides with the optical axis but points into the camera, in contrast to Tsai's setup.

3.9.6 Used Procedure

The absolute position of the calibration point is directly measured by the robot system by controlling the robot end effector to a known position and placing the calibration point at that position. This makes up the position of the calibration point ${}^{cp}_{abs}P$ expressed in the absolute coordinate frame. The calibration point is in fact the center of a small (black) circle.

Next, the robot end effector is controlled to a set of known positions ${}^{ee}_{abs}H(i)$, with ${}^{ee}_{abs}H$ the homogeneous representation[27] in absolute coordinates of the frame attached to the end effector and i the number of each position, going from 1 to e.g. 100. In order to get representative calibration parameters, the spread of the end effector positions covers the actual used workspace. For each of these positions the calibration point shifts relative to the end effector. The mounted camera, then, measures the centroid of the circle in the image plain, as the average position of the pixels belonging to the circle, in horizontal as well as in vertical direction.

This results, for each robot position, in the searched image point with sub-pixel accuracy, given by the set $[u(i), v(i)]$. See Fig. 3.36. The relative position of the calibration point w.r.t. the end effector ${}^{cp}_{ee}P(i)$ is computed for each robot position using the absolute measurements:

$$ {}^{cp}_{ee}\underline{P}(i) = [{}^{ee}_{abs}H(i)]^{-1}\ {}^{cp}_{abs}\underline{P} \tag{3.23}$$

with $\underline{P}$ the homogeneous coordinates $[x, y, z, 1]'$. If we split ${}^{ee}_{abs}H$ in a 3×3 rotation matrix and a 3×1 translation vector according to

$$ {}^{ee}_{abs}H = \begin{pmatrix} {}^{ee}_{abs}R & {}^{ee}_{abs}T \\ 0\,0\,0 & 1 \end{pmatrix} \tag{3.24}$$

then (3.23) becomes:

$$ {}^{cp}_{ee}P(i) = [{}^{ee}_{abs}R(i)]^{-1}\ ({}^{cp}_{abs}P - {}^{ee}_{abs}\,T(i)). \tag{3.25}$$

From (3.25) can be seen that if the orientation of the end effector (${}^{ee}_{abs}R$) does not change and if it is taken parallel to the base frame, the calculation of the relative position boils down to a simple subtraction, thus simplifying the method even more. Since both the end effector positions and the position of the calibration point are "measured by the robot", the method takes advantage of the *high relative positioning accuracy* of the robot.

The input for Tsai's calibration thus consists of the 3D (x, y, z) relative coordinates of the calibration point in mm, being ${}^{cp}_{ee}P(i)$, and the corresponding image coordinates $[u(i), v(i)]$ in pixels. Figure 3.36 gives an example. The external parameters $(\alpha_x, \alpha_y, \alpha_z, T_x, T_y, T_z)$, resulting form this calibration then determine the transformation from end effector to camera[28]:

$$ H_{cam-Tsai} = \begin{pmatrix} & T_x \\ R_z(\alpha_z).R_y(\alpha_y).R_x(\alpha_x) & T_y \\ & T_z \\ 0\ \ 0\ \ 0 & 1 \end{pmatrix} \tag{3.26}$$

[27] See also Appendix A, Sect. A.1.

[28] According to Tsai's setup, the ${}_{cam}z$-axis points out of the camera as shown in Fig. 3.34.

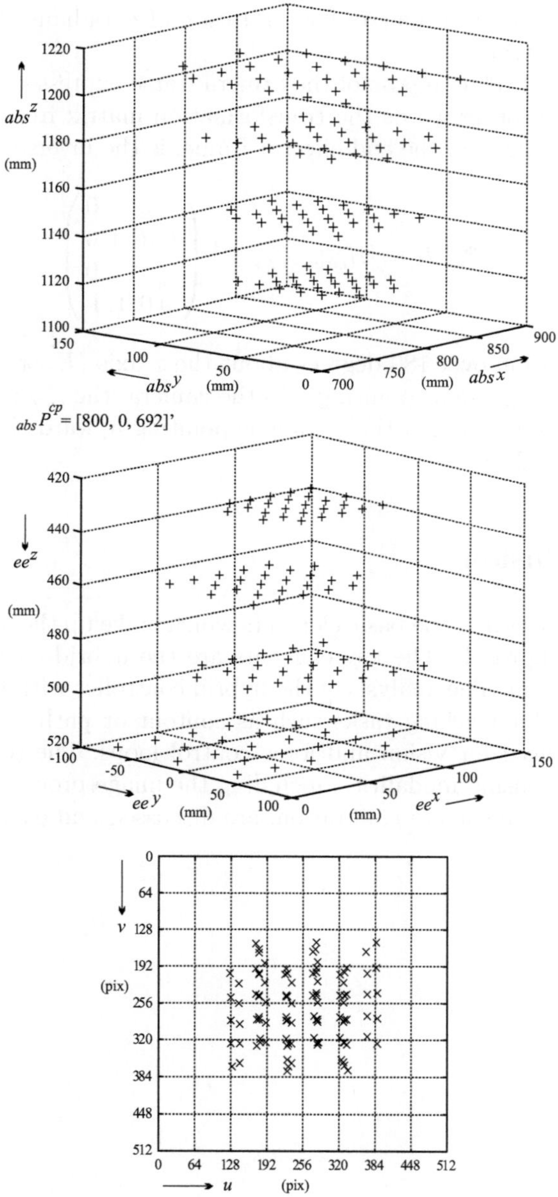

Fig. 3.36. Example of absolute end effector positions ${}^{ee}_{abs}P(i)$(**top**); resulting relative calibration point position in the end effector frame ${}^{cp}_{ee}P(i)$, which are used as input for the calibration (**middle**); and measured image plane coordinates $[u(i), v(i)]$ of the calibration point for the different end effector positions (**bottom**)

with $R_i(\alpha)$ the rotation matrix for a rotation of α radians about direction i (see also Sect. A.1).

The searched relative pose of the camera frame expressed in the end effector frame, or in other words the transformation matrix from camera coordinates to coordinates in the end effector frame, is the inverse of (3.26):

$$
{}^{cam}_{ee}H = \left[H_{cam-Tsai}\right]^{-1} \begin{pmatrix} & & & 0 \\ & R_x(\pi) & & 0 \\ & & & 0 \\ 0 & 0 & 0 & 1 \end{pmatrix} \tag{3.27}
$$

The extra rotation over 180 degrees about the x-axis ($R_x(\pi)$) gives a camera frame with the ${}_{cam}z$-axis pointing into the camera, the ${}_{cam}x$-axis pointing to the right of the image and the ${}_{cam}y$-axis pointing upwards in the image (see Fig. 3.33 and 3.35).

3.10 Conclusion

This chapter describes the basic elements which make up the underlying structure for our approach. The key elements are the hybrid control scheme and the TF formalism. The analysis of the hybrid controller with feedforward proves its fitness for tracking tasks such as contour or path following. Control gains are deduced for vision and force control loops. The (characteristics of the) separate sensing modalities, including the image processing, the contour modelling and the camera calibration, are discussed and explained.

4

Classification

4.1 Introduction

This chapter shows how to use the task frame to model, implement and execute 3D robotic tasks which combine force control and visual servoing in an uncalibrated workspace.

Chapter 3 explained the basics of the task frame formalism. It divides the control space, by means of the high level task description, in orthogonal directions as a first step to integrated vision/force control. Possible directions are *vision, force, tracking* and *velocity directions*, indicated by d^{vs}, d^{f}, d^{tr} and d^{v} respectively. The task description further specifies the set-points for force, velocity and vision directions. To any direction feedforward, indicated with $d^{+\mathrm{ff}}$, may be added, as shown in the control scheme of Fig. 3.1.

According to Nelson [60], the levels of force/vision integrated control are threefold, being *traded, hybrid* and *shared* control. In traded control, a given direction is alternately controlled by vision or by force. Hybrid control involves the simultaneous control of separate directions by vision and force. In shared control, which is the highest level of vision and force fusion, both sensors control the same direction simultaneously[1]. This chapter gives numerous examples of combined vision/force tasks, with the emphasis on hybrid and shared control.

The task frame concept provides a common frame to coordinate multiple external sensor-based controllers, especially when these sensors are mounted on the robot end effector. Mounting the camera on the end effector results in a controllable camera position. The image feature of interest is (mostly) placed

[1] A fourth vision/force sensing fusion occurs when the vision algorithm reconstructs the 3D position of a feature point, using the depth to the object from the 3D position of the force probe contact. This last method of sensor fusion depends on the relative mounting of the calibrated camera and the used vision algorithms.

on the optical axis (center of the image) by visual servoing[2], which makes the feature measurement less sensitive to calibration errors or distortions in the camera/lens system. Moreover, incorporating the vision information in the position control loop of the robot makes the control robust against kinematic or structural errors [44].

The combined mounting of force and vision sensors and the use of a control scheme with an internal velocity controller in combination with the task frame formalism, as shown in Fig. 3.1, distinguishes our approach from any other reported combined vision/force approach [43, 47, 60, 67, 95].

First, Sect. 4.2 summarizes the restrictions we impose on the considered tasks in a combined vision/force setup. This results in four (meaningful) distinctive configurations which are classified in Sect. 4.3. This classification depends on both the relative camera mounting and the involved control issues. Section 4.4 continues with the exploration of integrated vision/force tasks by giving all possible types of shared control in one direction. Next, Sect. 4.5 compares and discusses several full 3D tasks. Each task is explained by describing the control actions to be taken in the task frame. Finally, Sect. 4.6 concludes the chapter.

4.2 Adopted Restrictions

In robotic tasks, vision and force sensing are highly complementary. Handling soft objects, free space positioning, looking for a starting point, etc. are best handled by the vision system; making contact and tracking a surface in contact necessitate the use of a force sensor.

The force sensor gives full 3D information (6 force components) about the *local* contact with the object. Hence, the force sensor may control up to the full 6 DOF of the robot, depending on the actual contact situation between tool and environment.

The vision system, on the other hand, gives *global* information about the 3D environment projected onto the 2D image plane. Assuming the exact dimension or texture of the object or image feature is unknown[3], a distance measurement from image to object plane (or depth measurement) by the vision system is not possible[4]. Hence, the (mono) vision system can only measure, and therefore control, a maximum of 3 feature characteristics independently, being the perceived feature position (in x and y) and orientation (about z).

[2] According to the taxonomy introduced by [70] the used visual servoing type is a *dynamic look-and-move* control.

[3] This is the case for all considered tasks except for the visual alignment task of Chap. 5.

[4] Anyhow, if it were possible, such a depth measurement would be ill conditioned.

In order to 'fuse' the sensor signals, they are transformed or linked to the task frame[5]. A general transformation is very easy for the force measurements, but less so for visual information. After all, a general (3D) transformation of the vision measurement to the task frame would require the object depth and depth estimation from a single image is an ill conditioned operation. Therefore, the following discusses only those combined vision/force tasks in which the camera, if possible, is mounted in such a way that the vision controlled task frame directions lie either parallel (for a translation) or perpendicular (for a rotation) to the image plane. This simplifies the mapping (if needed!) from perceived image features (in the camera frame) to their equivalents in the task frame significantly. We refer to such a mounting as *parallel mounting*. Due to possible collision of the camera with the object, positioning the camera in a such way that the image plane lies parallel to the object may not always be feasible, necessitating a *non-parallel mounting*.

Furthermore, the tool or TF z-axis is by convention chosen identical (up to the sense) to the end effector z-axis. Hence, due to the physical connection between end effector and tool, the center of the camera frame can not lie on the tool z-axis[6].

Summarized, the adopted assumptions or constraints for the presented examples are:

1. Only tasks using combined mounting of vision and force sensor are considered.
2. Camera and TF z-axis can never coincide (geometric mounting constraint).
3. The camera frame may never collide with the object (geometric object constraint).
4. The distance between camera and object along the optical axis (z-axis) is not measurable by the vision system (observability constraint).
5. The optical axis points to the image feature of interest (resulting in minimal distortion) and, if possible, the image plane is taken parallel to the task frame (optimal camera positioning).

[5] Note that a transformation of the vision measurements to the TF is mandatory in position based visual servoing. In image based visual servoing, on the other hand, the (control of an) image feature is directly linked to a given TF direction. This is the case in the 3D alignment task of Chap. 5.

[6] Only in the special case of a force sensor with a hole in it and the camera looking through this hole, the given assumption is not valid. This special case is however not considered.

4.3 Tool/Camera Configurations

Two force tool/camera configurations are possible: either the camera sees the contact between tool and object or it does not see it.

ECL: If the camera observes both object and tool, the setup is referred to as *Endpoint Closed-Loop* (ECL). The visual servoing consist of controlling the relative position of the force probe w.r.t. the object as seen by the camera. In this setup, the optical axis of the camera points towards the center of the task frame or to the tool center point (TCP). Both measurement and control by vision as well as by force are collocated at the same point, i.e. the TF origin. An ECL setup may further be divided into two cases depending on the object constraint and the optimal camera positioning.

Parallel vs. Non-parallel ECL: If feasible, the image plane is taken parallel to the TF. Otherwise, a non-parallel setup is chosen. The constraining factor in the above choice is of course the fact that object and camera may never collide. Figure 4.6 gives examples of ECL configurations with parallel (4.6-top) and non-parallel mounting (4.6-bottom).

An ECL visual servoing action is possible without calibration, although the exact relation between the controlled velocity and the change in measured feature error is, with an uncalibrated camera, unknown and can therefore not be taken into account in the control law. If, on the other hand, camera pose and force tool are calibrated, the exact 3D positions and errors of the contact point or image feature are known, making a more accurate control design possible.

EOL: If, on the other hand, the camera does not see the tool, the setup is referred to as *Endpoint Open-Loop* (EOL). In contrast to the ECL case, an EOL setup comes down to a non-collocated control. Not only the actual TF pose, but also the pose of the camera frame have to be controlled.

Again two subcases are possible: either the relation of the camera frame to the task frame is *fixed* or it is *variable.* Both cases use a parallel mounting. The difference between the two configurations depends on the existence (and use) of a redundancy for the TF orientation. This redundancy is explained first. Then the fixed and variable EOL configurations are discussed.

Redundancy for TF orientation: Since task and end effector frames are not necessarily rigidly fixed to each other, the TF orientation possesses a redundant level of control, if a rotationally symmetric tool is used. For example, reorienting the task frame can be done (*i*) by rotating the robot end effector while keeping the task frame fixed to the end effector or (*ii*) by redefining the relation, in particular the angle about the z-axis, between task and end effector frames. Hence, an additional requirement may be imposed on the TF/end effector orientation control.

Figure 4.1 gives three examples applied to a contour following task. The task objective consists of keeping the task frame tangent to the contour. However, the way this is realized differs from case to case. In the first case (left), the task frame is kept tangent to the contour (while moving along the contour) by rotating both task frame and end effector. The task frame is *fixed* to the end effector. In the second case (middle) the task frame is kept tangent to the contour by redefining the relation between task frame and end effector. The orientation of the end effector does not change but the task frame can *rotate* with respect of the end effector. In the third case (right), the angle between task and end effector frames is again variable but in addition the end effector orientation may change. The variable orientation of the end effector, which is independent of the TF orientation, can now be used to position the end effector mounted camera[7].

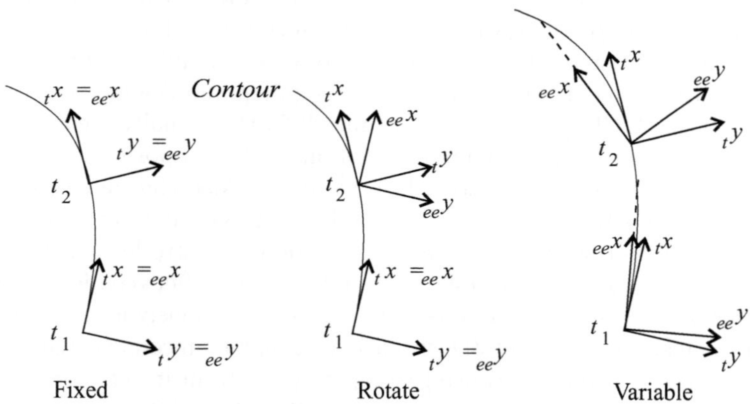

Fig. 4.1. Examples for two time instants (t_1 and t_2) of a TF *fixed* to the end effector (**left**), a *rotating* TF with fixed end effector orientation (**middle**) and both *variable* task and end effector orientations (**right**)

Fixed EOL: Since the camera is rigidly mounted on the end effector, the camera frame is always fixed w.r.t. the end effector (frame). Hence, the fixed case of Fig. 4.1 also corresponds to a fixed camera-task relation. Here, some TF directions are used to control the pose of the camera frame, in such a way that the image errors[8] diminish (to zero). However, controlling the camera pose will influence the TF or tool pose as well, since both are fixed w.r.t. each other.

With image based (fixed EOL) visual servoing, there is even no need for the mentioned general transformation of the vision measurements to their equivalents in the task frame. Nor a calibration of the camera setup is neces-

[7] This is yet another extension to the task frame formalism, similar to those presented by Qui Wang [91].

[8] i.e. the difference between desired and measured image features.

sary. However, as is the case in the ECL configuration, a calibrated camera is advantageous for the control design.

Variable EOL: In the variable camera frame/task frame case, the pose of the camera may change without actually changing the TF pose. This corresponds to the variable case of Fig. 4.1 using the redundancy which exists for rotationally symmetric tools. If the tool axis is identical to the last revolute joint axis, rotating the end effector changes the position of the eccentric mounted camera, while the tool center itself stands still. During this motion, the TF pose stays fixed w.r.t. any absolute frame, but shows a varying rotation about the $_tz$-axis w.r.t. the end effector.

An example of the *variable* camera-task relation case in the EOL configuration is the planar combined vision/force contour following task described in Chap. 6. In this task, the camera looks ahead, while moving along the contour, to measure (and log) in advance the path which the tool has to follow later on. Hence, in contrast to the fixed case, there is a (possibly variable) time shift between the moment the vision data are acquired and the moment these data are used (in the TF). To solve this time shift, the visually measured path is linked to the actual contact point based on Cartesian positions. For a planar task, this linking method is straightforward, provided the camera setup is calibrated, as explained in Chap. 6. For a 3D task, on the other hand, the exact depth of the feature at the moment the vision data are logged, is unknown. But the depth to the object follows from the contact between force probe and object when the force probe passes through the previously logged position. At this time instant, the exact 3D (visual) error may be reconstructed in order to compute the appropriate control or feedforward from it. This, in contrast to any other configuration, necessitates the calibration of the camera pose w.r.t. the end effector.

Advantages and Disadvantages: The main advantage of an ECL configuration is the trivial fact that the tool always lies in the camera field of view. The main disadvantage may be the *occurrence of occlusion* of the object by the tool, making a straightforward vision measurement impossible. Furthermore, compensation for the vision processing delay is not possible. A disadvantage specific to a non-parallel ECL configuration is the inaccurate (or unknown) mapping of vision measurements to the task frame.

The main advantage of an EOL configuration is the *absence of occlusion* of the object by the tool. Hence, easy image processing algorithms can be used. Furthermore by looking ahead, image capture and control delays can be compensated. The main disadvantages are the needed link, if image data are to be used in the task frame, and the extra control involved in the variable camera frame/task frame relation. Both disadvantages are, however, adequately solved in Chap. 6.

As mentioned, calibration of the camera setup is essential in a variable EOL configuration in order to link the vision information to the contact at hand[9]. However, for all configurations, a calibrated camera improves the control design.

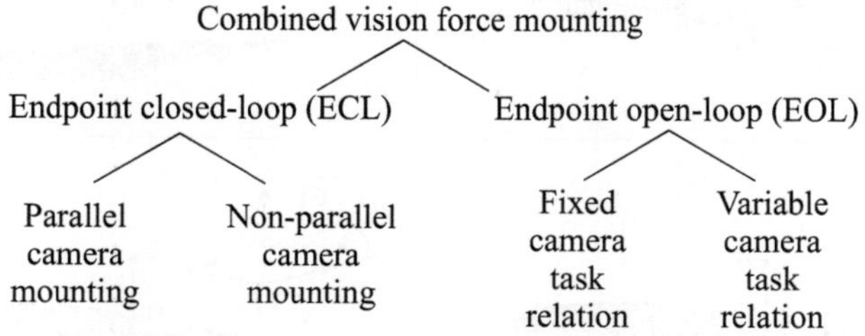

Fig. 4.2. Classification of combined vison/force configurations

Summary: This section gives four meaningful tool/camera configurations. They are on the one hand the endpoint closed-loop configurations with either a parallel or a non-parallel camera mounting and on the other hand the endpoint open-loop configurations with either a fixed or a variable camera frame to task frame relation. Figure 4.2 gives an overview.

4.4 Shared Control Types

The highest level of sensor integration is shared control. There are six (meaningful) basic forms of shared control: three for an axial direction (a) and three (counterparts) for a polar direction (p). Figure 4.3 shows one example for each possible form of shared control:

[a1,p1:] *Vision and force control on the same direction* (d^{f+vs}): The objective of the approach task (a1) of Fig. 4.3 consists of establishing a point/plane contact. The shared control action yields

$$ {}_tv_z^c = (K_z^f({}_tF_z^m - {}_tF_z^d)/k_z^t) + (K_z^{vs}\Delta z). \tag{4.1}$$

The first part of (4.1) is force dependent. The second part represents the visual servoing action. Both parts correspond to the proportional control laws of table 3.1. The force dependent part of the approach velocity ${}_tv_z^c$ is, in free space, equal to $-K_z^f {}_tF_z^d/k_z^t$ with ${}_tF_z^d$ the desired contact force, K_z^f the control gain and k_z^t the estimated tool stiffness. To avoid excessive contact forces upon

[9] Due to the calibration, the vision algorithms too will benefit from the combined vision/force setup. If the relative position of camera and force sensor is known and if the (calibrated) tool is in contact with the object, the depth from the image to the object plane is known. Hence, the 3D position of the image feature can be calculated easily (with mono vision).

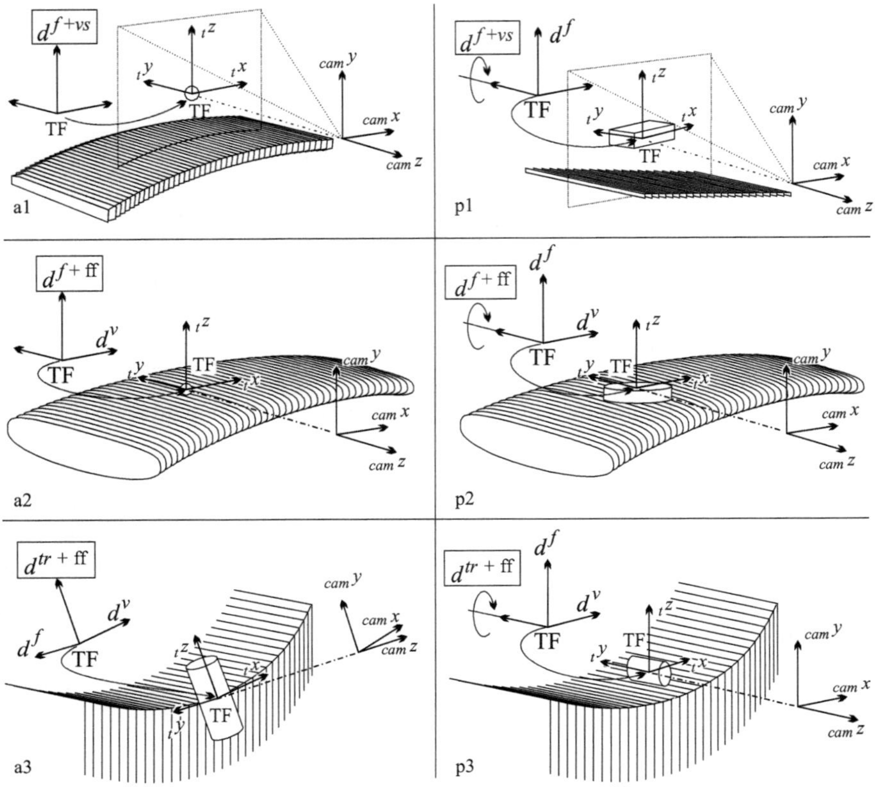

Fig. 4.3. Task examples of the six possible shared control forms: 3 axial (**left**) and 3 polar (**right**)

contact, this force dependent part cannot be too large. Hence, a small control gain and therefore a slow approach velocity is mandatory if only force sensing were to be used. The vision sensor, on the other hand, measures the distance between tool and plane Δz, proportional to which the approach velocity is increased. This accelerates the execution of the task.

In case (p1) the task objective is to make a line/plane contact. The desired torque ${}_tM_y^d$ is zero. The torque control on the ${}_ty$-direction will only upon contact align the tool with the plane[10]. The vision system, on the other hand, measures the angle $\Delta\theta_y$ between tool and plane in order to align the tool with the plane (possibly even before contact occurs). The corresponding shared control yields

$$ {}_t\omega_y^c = (K_{\theta y}^f({}_tM_y^m - {}_tM_y^d)/k_{\theta y}^t) + (K_{\theta y}^{vs}\Delta\theta_y). \tag{4.2}$$

[10] The tool will move towards the object due to the force control in the ${}_tz$-direction.

In this example, additional shared control of type (a1) can be applied in ${}_tz$-direction.

[a2,p2:] *Force control with vision based feedforward* ($d^{f+\mathrm{ff}}$): Examples (a2) and (p2) illustrate tasks using shared control in order to maintain a point/plane and line/plane contact respectively, while moving along the plane. In task (a2) the task frame specification does not allow a rotation around the ${}_ty$-axis. Hence, the ${}_tz$-direction is fixed. The shared control in the ${}_tz$-direction yields

$$ {}_tv_z^c = (K_z^f({}_tF_z^m - {}_tF_z^d)/k_z^t) + ({}_tv_x^c \, \tan({}_t\theta_y)), \tag{4.3} $$

with ${}_t\theta_y$ the angle of (the tangent to) the contour to be followed. In this case, the feedforward component (${}_tv_x^{\mathrm{ff}} = {}_tv_x^c \, \tan({}_t\theta_y)$) is tangent based.

In task (p2), the ${}_tz$-direction has to be kept normal to the plane. This is accomplished by demanding a zero torque around the ${}_ty$-direction (${}_tM_y^d = 0$). The shared control for the rotation around the ${}_ty$-direction yields

$$ {}_t\omega_y^c = (K_{\theta y}^f({}_tM_y^m - {}_tM_y^d)/k_{\theta y}^t) + ({}_tv_x^c \, \kappa_y), \tag{4.4} $$

with κ_y the curvature of the contour or path to be followed in the xz-plane. Hence, in case (p2) the feedforward component (${}_t\omega^{+\mathrm{ff}} = {}_tv_x^c \, \kappa_y$) is curvature based. An additional shared control in the ${}_tz$-direction of type (a2) is possible.

[a3,p3:] *Tracking control with vision based feedforward* ($d^{tr+\mathrm{ff}}$): In example (a3) the tool is in contact with an edge. While moving along the edge the tracking action (d^{tr}) keeps the contact point centered, i.e. in the middle of the tool resulting in zero torque around the ${}_tx$-axis. The shared control yields

$$ {}_tv_z^c = (K_z^{tr} \, \Delta z^{tr}) + ({}_tv_x^c \, \tan({}_t\theta_y)), \tag{4.5} $$

with Δz^{tr} the tracking error, which is in this case identified from

$$ \Delta z^{tr} = -{}_tM_x^m / {}_tF_y^m. \tag{4.6} $$

Again, the vision based feedforward component, which is the latter part of (4.5), is tangent based. It keeps the tool centered with respect to the contour when moving along the contour, once this centered position is reached. This reduces the (occurrence of) tracking errors significantly.

Example (p3) illustrates a task in which a line/plane contact has to be maintained while moving along an one-dimensionally curved plane. The tracking control has to keep the forward direction (d^v) tangent to the plane by rotating around the ${}_ty$-axis. The shared control for the ${}_ty$-rotation yields

$$ {}_t\omega_y^c = (K_{\theta y}^{tr} \, \Delta\theta_y^{tr}) + ({}_tv_x^c \, \kappa_y), \tag{4.7} $$

The vision based feedforward ${}_t\omega^{+\mathrm{ff}} = {}_tv_x^c \, \kappa_y$ is proportional to the curvature κ_y and will reduce the tracking error $\Delta\theta_y^{tr}$.

Any of the given tasks can be executed without using vision. However, either the task quality (with regard to tracking errors or acting contact forces) or the execution time needed will be worse in the force-only case than in the shared control case.

All the examples of Fig. 4.3 use an ECL configuration, with parallel camera mounting. Counterparts for non-parallel mounting or EOL systems exist.

Notice that there are no examples given with tracking and visual servoing on the same direction (d^{tr+vs}). This combination, however possible, is inferior to the case of tracking with vision based feedforward (cases a3 and p3 in Fig. 4.3). If shared tracking/visual servoing is feasible, tracking with vision based feedforward will be feasible too and will result in a better control.

Also visual servoing with feedforward is possible ($d^{vs+\text{ff}}$). This, however, involves no shared control.

Any task can further be classified according to the number of shared and/or vision controlled directions. Assuming the object dimension is unknown, the (mono or uncalibrated) vision system can only control a maximum of three directions independently. Hence, a given task can only possess a maximum of three shared controlled directions. Possible combinations (l, k), with (l) the number of shared controlled directions and (k) the number of vision(-only) controlled directions, fulfilling the condition $l + k \leq 3$, are: (1,0), (1,1), (1,2), (2,0), (2,1), (3,0). This gives again six different combinations or classes of tasks containing at least one direction with shared control. The next section gives some examples.

A further distinction in the task may follow from the way the remaining directions are divided into force, tracking and velocity directions. Such a subdivision, however, would leads us too far.

4.5 Combined Vision/Force Task Examples

This section gives several combined vision/force task examples which fulfill the assumptions and constraints given in Sect. 4.2.

First, some traded control examples are illustrated. Traded control is mostly used in one or more (distinctive) subtasks, which together make up the global task. Note that the actual location of the task frame may change from subtask to subtask and may as well coincide with the camera frame.

Then, a fairly extensive list of task examples using hybrid and/or shared control is given. This list is not intended to be exhaustive. It merely illustrates the vast set of tasks in which meaningful shared and hybrid control may be used. All figures show the chosen control types for any direction (vision vs, force f or velocity v possibly with feedforward +ff). If nothing is indicated, the concerned direction is velocity controlled with desired velocity set equal to zero.

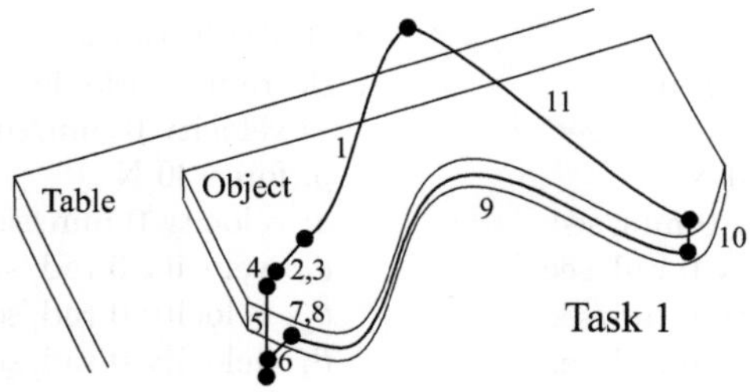

Fig. 4.4. Paths travelled according to the different subtasks in order to automatically search a starting point, follow the contour and go home

For some tasks, also the high level task description program is given. The syntax and units used in this specification automatically indicate the control type.

4.5.1 Traded Control Example

Consider the global task, shown in Fig. 4.4, which consists of following subtasks: 1.1) go to a fixed position, 1.2) search a starting point on the edge by vision, 1.3) align with the contour tangent using vision, 1.4) move to a safe approach position (otherwise the force probe would hit the object on the top plane instead of the table for the given setup), 1.5) make a force controlled contact with the table (establishing the depth), 1.6) lift the robot tool to a safe height above the table, 1.7) make contact with the edge of the object, 1.8) position the camera over the contour while maintaining the force control, 1.9) move along the contour with a given tangential velocity, while maintaining a constant normal contact force, adjusting the TF directions by tracking and feedforward and keeping the contour in the camera field-of-view until a given distance is travelled [11] 1.10) retract the robot tool from the contour surface and 1.11) go to the home position.

SUBTASK 1.3: {
task frame: fixed
  x: velocity 0 mm/sec
  y: feature distance in camera frame 0 mm
  z: velocity 0 mm/sec
  θ_x: velocity 0 rad/sec
  θ_y: velocity 0 rad/sec
  θ_z: feature angle in camera frame 0 rad
until θ_z feature angle >-0.1 rad and <0.1 rad
  and y feature distance <0.5 mm and >-0.5 mm }

[11] Subtask 1.9, the contour following part, is the subject of Chap. 6 and corresponds to Task 2b in Fig. 4.5-bottom.

SUBTASK 1.7: {
task frame: fixed
x: velocity 0 mm/sec
y: force 20 N
z: velocity 0 mm/sec
θ_x: velocity 0 rad/sec
θ_y: velocity 0 rad/sec
θ_z: velocity 0 rad/sec
until y force > 15 N }

SUBTASK 1.8: {
task frame: variable
x: velocity 0 mm/sec
y: force 30 N
z: velocity 0 mm/sec
θ_x: velocity 0 rad/sec
θ_y: velocity 0 rad/sec
θ_z: velocity 0 rad/sec
until y feature distance in camera frame
< 0.1 mm and > -0.1 mm }

Subtask 1.9 corresponds to the combined vision/force contour following task discussed in Chap. 6. Subtasks 1.3, 1.7 and 1.8, for which experimental results are given in Chap. 8, are programmed as shown above. Subtasks 1.3 and 1.7 are an example of traded control. In subtask 1.3 the $_ty$-direction is vision controlled, in subtask 1.7 the $_ty$-direction is force controlled.

4.5.2 Hybrid and Shared Control Examples

This paragraph gives several task examples using hybrid and/or shared control. For reasons of convenience the tasks are numbered. This numbering is merely for referencing purposes and does not imply a classification or degree of difficulty. In each figure the Control Space Dimension (CSD) is given. CSD is the number of independent states to be controlled in the task. If a variable camera frame is used, the orientation control of this frame introduces one extra

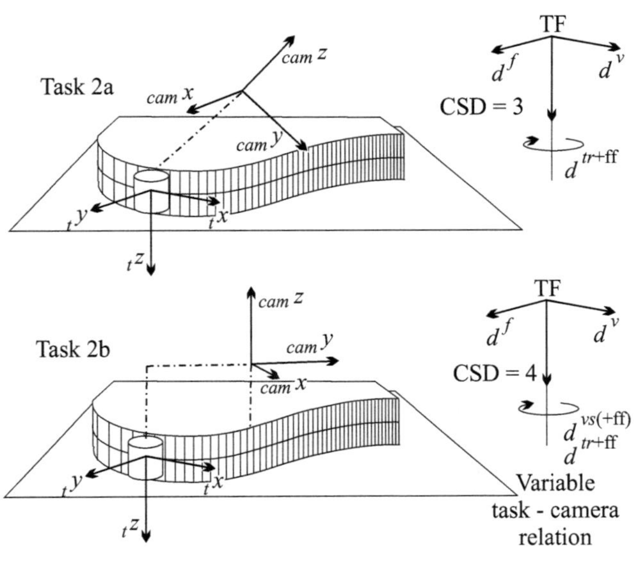

Fig. 4.5. Planar contour following tasks in a non-parallel ECL configuration (**top**) and in a variable EOL configuration (**bottom**)

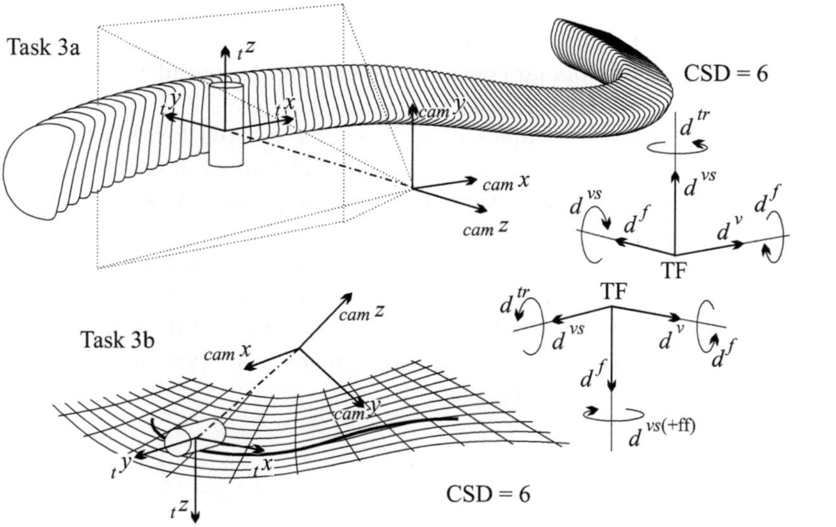

Fig. 4.6. Pipe (**top**) and path on surface (**bottom**) following tasks in ECL configurations

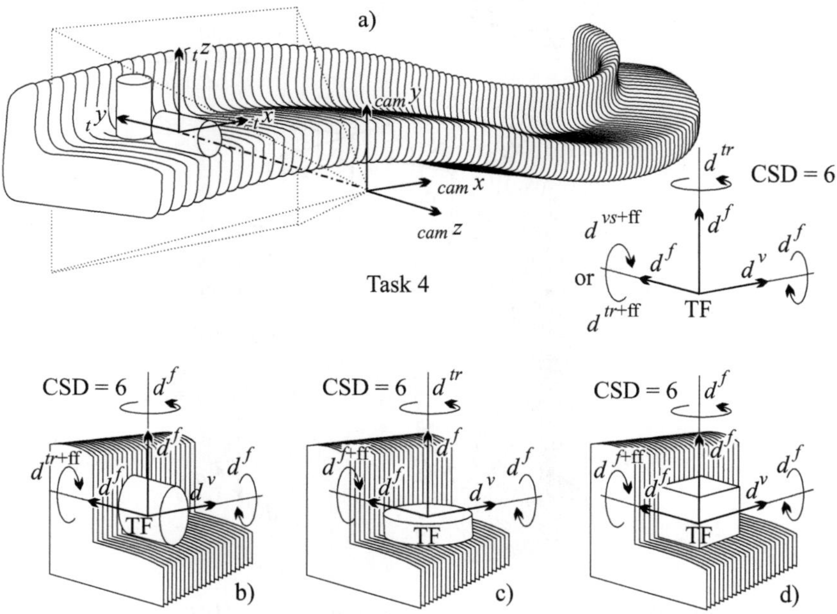

Fig. 4.7. Seam tracking tasks in a parallel ECL configuration

control state. Hence the maximum control space dimension is 7. Some tasks only differ in the tool/object contact situation. At the end of this section, table 4.1 compares and summarizes all the characteristics of the presented hybrid or shared controlled tasks.

Figure 4.5 shows planar contour following in non-parallel ECL and variable EOL.

Figure 4.6 gives examples of ECL configurations for which the CSD is equal to 6, however, without shared control. The pipe following task of Fig. 4.6 (top) allows a parallel mounting. The path on plain task of Fig. 4.6 (bottom) necessitates a non-parallel mounting.

Figure 4.7 gives four examples of seam tracking tasks, each of which has one shared controlled direction and one velocity direction. Depending on the tool, the other directions are divided into force and tracking directions. Remember that a tracking direction uses measurements from two other directions to automatically adjust the TF pose.

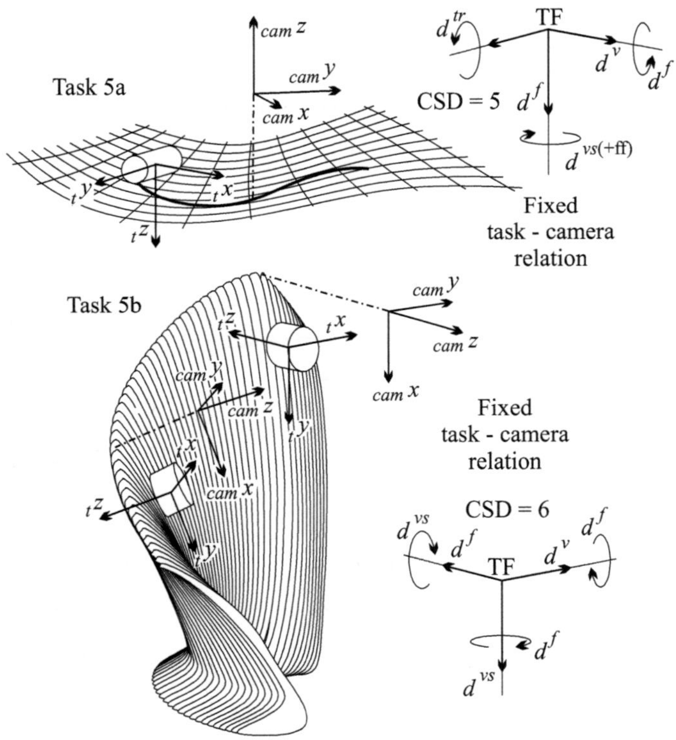

Fig. 4.8. Examples of fixed EOL tasks: path on surface (**top**) and blade polishing (**bottom**)

Figures 4.8 and 4.9 give examples of EOL configurations with control space dimensions ranging from 5 to 7. The path on surface following task of Fig. 4.8 (top) with CSD = 5, can be compared to a truck with semi-trailer. The camera is the 'truck', the tool the 'trailer'. The optical axis of the camera always stays on the path. The relation between task and camera frame is fixed, hereby setting the direction in which the trailer goes.

Also the blade polishing task of Fig. 4.8, has a fixed task to camera relation. Here the dimension of the control space is equal to 6. The blade polishing task of Fig. 4.8 (bottom) consists of following at a safe distance the edge of a slightly curved surface with a polishing tool while maintaining a constant contact force. The tool must be kept normal to the surface. The task program is:

TASK 5b - 3D blade polishing – edge following: {
with task frame: fixed
  x: velocity 10 mm/sec
  y: feature distance in camera frame 0 mm
  z: force -15 N
  θ_x: force 0 Nmm
  θ_y: force 0 Nmm
  θ_z: feature angle in camera frame 0 rad
until relative distance > 250 mm}

In this example, the ${}_tz$-, the ${}_t\theta_x$- and the ${}_t\theta_y$-directions are force controlled; ${}_tx$ is a velocity direction, ${}_ty$ and ${}_t\theta_z$ are vision directions. The desired contact force in the ${}_tz$-direction $= -15$ N. Chapter 8 gives experimental results for this type of task.

The examples of Fig. 4.9 show a variable EOL seam tracking task and a variable EOL path on surface following task. Both tasks have a control space dimension of 7. They can be seen as the 3D extension of the contour following tasks of Fig. 4.5 (bottom). A task description for the path on surface task of Fig. 4.9 (top) is:

TASK 6a - 3D path on surface following: {
with task frame: variable
  x: velocity 20 mm/sec
  y: feature distance 0 mm
  z: force -10 N
  θ_x: force 0 Nmm
  θ_y: force 0 Nmm
  θ_z: feature angle 0 rad with feedforward
until relative distance > 250 mm}

Finally, Fig. 4.10 shows a block on plane positioning task with three vision controlled directions and Fig. 4.11 gives a block in corner positioning task with the maximum of three shared controlled directions.

To conclude this section, the comparative table 4.1 gives an overview of the presented tasks. Experimental results for some of these tasks are presented in Chaps. 6 and 8. Chapter 6 fully explores the planar contour following task (Task 2b in Fig. 4.5-bottom) which uses a variable EOL configuration. Chapter 8 discusses the experimental results for Task 1 (subtasks 1.3, 1.7 and 1.8)

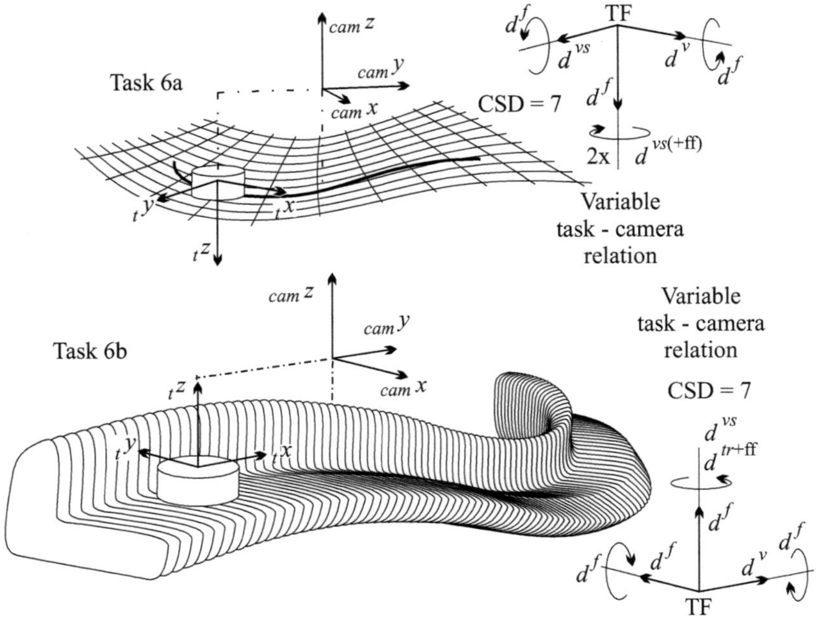

Fig. 4.9. Examples of variable EOL tasks: path on surface (**top**) and seam tracking (**bottom**)

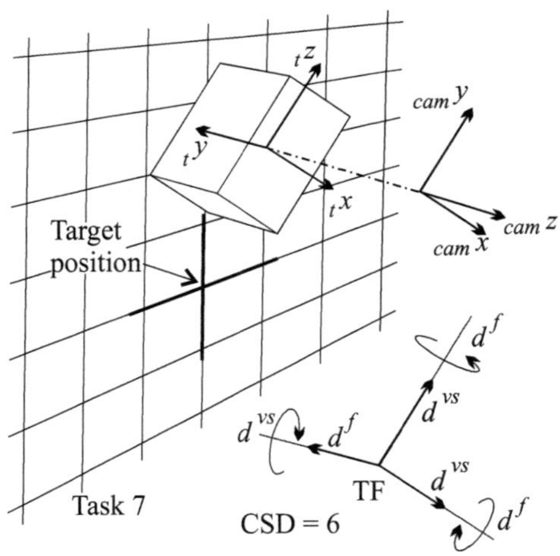

Fig. 4.10. Block on plane positioning task

as an example of traded control. It continues with the results of a shared control task of type 1a (according to Fig. 4.3) and concludes with the blade polishing experiment Task 5b (Fig. 4.8-bottom).

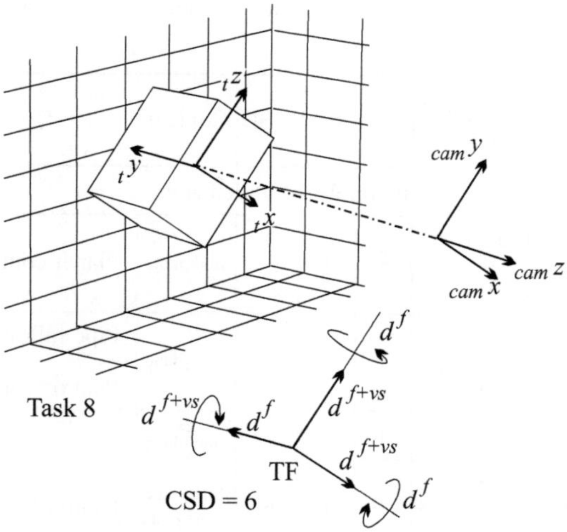

Fig. 4.11. Block in corner positioning task

4.6 Conclusion

This chapter shows how the task frame formalism provides the means to easily model, implement and execute 3D robotic servoing tasks using both force control and visual servoing in an uncalibrated workspace.

A framework is established to classify and compare combined vision/force tasks. On the one hand, four meaningful camera/tool configurations are suggested, of which the endpoint open-loop configuration with variable task/camera frame relation is the most challenging in the sense of control. On the other hand, possible types of shared control, which is the highest level of vision/force integration, are discussed and illustrated.

Numerous tasks exemplify the four camera/tool configurations as well as traded, hybrid and shared vision/force control. Together with the presented high level task descriptions, they emphasize the potential of the used approach to widen the range of feasible tasks.

Table 4.1. Overview of the characteristics of tasks 1 to 8, presented in this section in Figs. 4.4 to 4.11

<table>
<tr><th>Task no</th><th>Con-fig.</th><th>$\#d^{vs}$ (*)</th><th>$\#d^{f}$</th><th>$\#d^{tr}$</th><th>$\#d^{v}$</th><th>CSD</th><th>Contact</th><th>Description</th></tr>
<tr><td>1</td><td colspan="8">Example of traded control</td></tr>
<tr><td>2a</td><td>X</td><td>(1,0)</td><td rowspan="2">1</td><td rowspan="2">-</td><td rowspan="2">1</td><td>3</td><td rowspan="2">line/plane</td><td rowspan="2">Planar contour following</td></tr>
<tr><td>2b</td><td>Var.</td><td>(1,0)</td><td>3+1</td></tr>
<tr><td>3a</td><td>//</td><td>(0,2)</td><td rowspan="2">2</td><td rowspan="2">1</td><td rowspan="2">1</td><td rowspan="2">6</td><td rowspan="2">line/plane</td><td>Pipe following</td></tr>
<tr><td>3b</td><td>X</td><td>(0,2)</td><td>Path on surface</td></tr>
<tr><td>4a</td><td rowspan="4">//</td><td>(0,1)
(1,0)</td><td>3</td><td>1</td><td rowspan="4">1</td><td rowspan="4">6</td><td>2x
line/plane</td><td rowspan="4">Seam tracking</td></tr>
<tr><td>4b</td><td>(1,0)</td><td>4</td><td>-</td><td rowspan="2">line/plane
plane/plane</td></tr>
<tr><td>4c</td><td>(1,0)</td><td>3</td><td>1</td></tr>
<tr><td>4d</td><td>(1,0)</td><td>4</td><td>-</td><td>2x
plane/plane</td></tr>
<tr><td>5a</td><td rowspan="2">Fix.</td><td>(0,1)</td><td>2</td><td>1</td><td>1</td><td>5</td><td>line/plane</td><td>Path on surface</td></tr>
<tr><td>5b</td><td>(0,2)</td><td>2
3</td><td>1
-</td><td>1</td><td>6</td><td>line/plane
plane/plane</td><td>Blade polishing</td></tr>
<tr><td>6a</td><td rowspan="2">Var.</td><td>(0,2)</td><td>3</td><td rowspan="2">-</td><td rowspan="2">1</td><td rowspan="2">6+1</td><td>plane/plane</td><td>Path on surface</td></tr>
<tr><td>6b</td><td>(1,0)</td><td>4</td><td>line/plane
plane/plane</td><td>Seam tracking</td></tr>
<tr><td>7</td><td>//</td><td>(0,3)</td><td>3</td><td>-</td><td>-</td><td>6</td><td>plane/plane</td><td>Block on plane</td></tr>
<tr><td>8</td><td>//</td><td>(3,0)</td><td>3</td><td>-</td><td>-</td><td>6</td><td>3x
plane/plane</td><td>Block in corner</td></tr>
</table>

Legend:

#: number of
(*) (#shared, #vision)
EOL: Endpoint open loop
ECL: Endpoint closed loop
CSD: Control space dimension

// Parallel camera mounting, ECL
X Non-parallel camera mounting, ECL
Var: Variable camera task relation, EOL
Fix: Fixed camera task relation, EOL

5

Visual Servoing: A 3D Alignment Task

From a historical point of view, the visual servoing related aspects of this work are the main new contributions to the already implemented [83] force controlled robot structure[1]. To illustrate the visual servoing capabilities (apart from any force controlled task), the following exemplifies a strategy to visually position the robot (with end effector mounted camera) relative to an arbitrarily placed object. Such an alignment task may be part of a more global task consisting of first locating the object, then aligning with the object and finally grasping the object. The (initial) 3D pose of the object w.r.t. the robot is of course unknown. This chapter also investigates the suitability of three kinds of reference objects or markers[2]: a circle, an ellipse and a rectangle.

The 3D relative positioning task is solved using image based visual servoing, meaning that the control signals are computed directly from the image features. The approach is, among others, based on the relative area method, explained in Sect. 3.6. It uses a coarsely known camera setup. In particular, a rough estimation of the focal length and the orientation of the eye-in-hand camera w.r.t. the end effector are sufficient.

5.1 Strategy

A 3D positioning task involves the control of the robot for the full 6 DOF. In a visual alignment task this boils down to controlling the position and orientation of the camera relative to a given object (using vision information only). For this task, the task frame is centered at the camera and is in fact identical

[1] at the department of Mechanical Engineering of the Katholieke Universiteit Leuven

[2] The reference marker may be placed on the object or may be the object itself. In the following the reference 'object' used for the alignment task is referred to as *alignment object.*

to the camera frame[3]. For reasons of simplicity, the preceding subscripts t or cam, indicating task or camera frame, which are in this task identical, and the preceding superscript obj, indicating a parameter of the object, are (mostly) omitted in the following.

Each task direction is controlled continuously using specific vision information or image features. The alignment object, which still may be chosen freely, needs to posess - and preferably emphasize - the desired vision information. The (desired) end pose of the robot (i.e. the relative pose of the robot w.r.t. the object at the end of the task) has to follow unambiguously from the pose of the alignment object. Possible alignment objects are a circle [47, 71], an ellipse [33], a rectangle, a set of points [14, 34] or any combination of these. The following investigates the use of a circle, an ellipse and a rectangle. The latter is found to be the best choice.

5.1.1 3D Alignment Using a Circle

A circle alignment task may be described as follows:

> *Position the camera in such a way that (i) the center of the circle lies at the optical axis (using 2 DOF), that (ii) the image plane lies parallel to the object plane (using again 2 DOF) and that (iii) the distance between the object plane and the image plane amounts to a given value (using 1 DOF).*

The task description specifies 5 of the 6 degrees of freedom. The 6th DOF, being the rotation around the optical axis, can not be defined since a circle is rotationally symmetric. For the distance control either the real length of the object (mm) (e.g. the diameter) or the desired perceived object length in the image plane (pix) must be known.

Figure 5.1 shows the four steps in realizing the given task. The shown velocities indicate the control actions to be taken. The four steps are:

1. Center the perceived image by translating in x and y-direction[4].
2. Align the longest axis of the circle, which is perceived as an ellipse if image plane and object plane are not parallel, with the x-direction by rotating around the z-axis. Determine the distance to the center of the circle, measured along the optical axis as shown in Fig. 5.2, by measuring the width of the circle (i.e. the longest axis of the perceived ellipse which now lies horizontally).
3. Maximize the measured object size along the y-axis, by rotating around the x-axis. Simultaneously translate along the y-direction, so that the actual followed camera path is arc shaped with the center of this path at

[3] For a typical setup of the camera and the definition of the camera frame, see Fig. 3.3.

[4] Mark that the center of the perceived ellipse is not the center of the real ellipse.

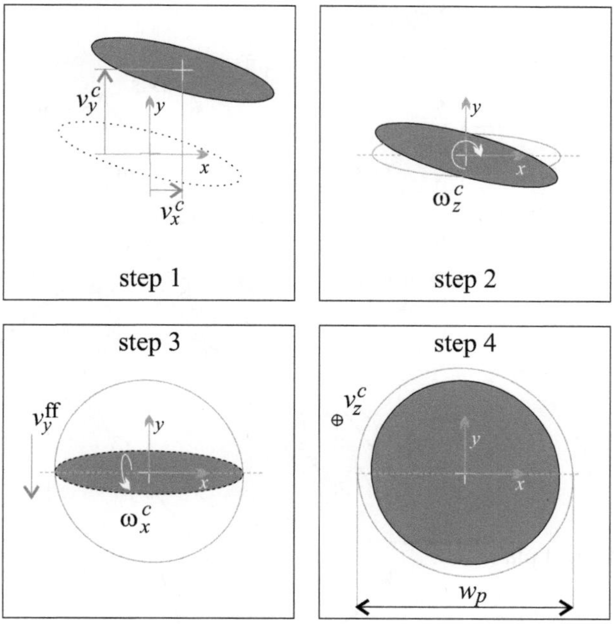

Fig. 5.1. Four steps of the (eye-in-hand) circle alignment task

the object. See Fig. 5.2. The perceived image will slowly evolve from an ellipse to a circle. The image plane is now parallel to the object plane
4. Adjust the distance between the image plane and the object plane by translating over the z-axis.

With each new step, the control actions for the previous steps are maintained. The execution of a new step starts if the end condition of the previous step is fulfilled. A typical end condition is 'perceived error smaller than a given threshold'.

The main advantage in using a circle as alignment object, is the fact that, provided the real diameter of the circle is known, the distance to the object plane is derived from the longest perceived axis independent from the camera pose.

However, using a circle also causes several problems. As indicated in Fig. 5.3, the sign of the action to be taken in step 3 does not unambiguously follow from one image. Both camera positions a and b give (almost) the same perceived image. The correct sense of the control action to be taken follows only from a sequence of images (in motion) by checking whether the measured object size along the y-axis, defined as the object height, increases. If the perceived object height increases, the current sense of the motion is correct. If the perceived object height decreases, the current motion sense has to be inverted. Due to measurement noise, the controller must not react too quickly to a change in measured object height. A filter or non-linear elements with

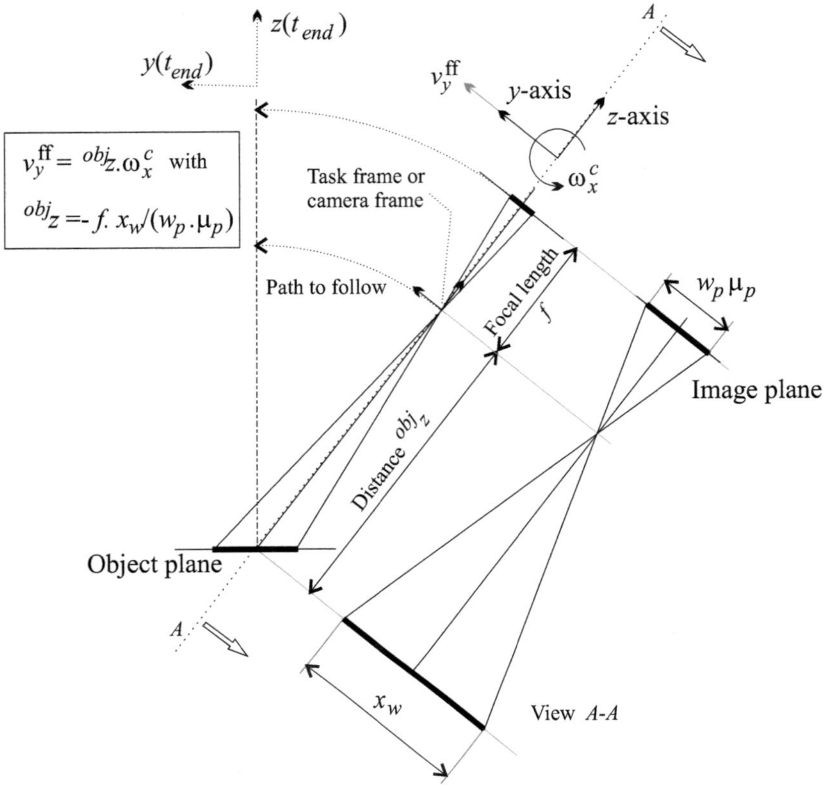

Fig. 5.2. Determination of v_y^{ff} (mm/s) in order to follow an arc shaped path around the center of the object, given the object width x_w (mm), the measured width in pixels w_p, the pixel size μ_p (mm/pix), the focal length f (mm) and the commanded polar velocity ω_x^c (rad/s)

hysteresis can partly stabilize the measurement signal. They, however, also introduce phase lag in the measurement and may affect the accuracy of the positioning or the stability of the control. The biggest problems in determining the correct sense of the motion (in step 3) occur when the actual camera position is close to the desired position. Here, image and object planes are almost parallel and a rotation around the x-axis will influence the measured object height only slightly. The positioning task using a circle as alignment object is thus ill conditioned.

Moreover, experiments show that also step 2 causes problems. When image and object planes lie parallel - which is the goal of the task -, the determination of the longest axis of the perceived circle is very difficult and not clearly defined. Hence, the action which is based on this determination, has to be performed very slowly and accurately.

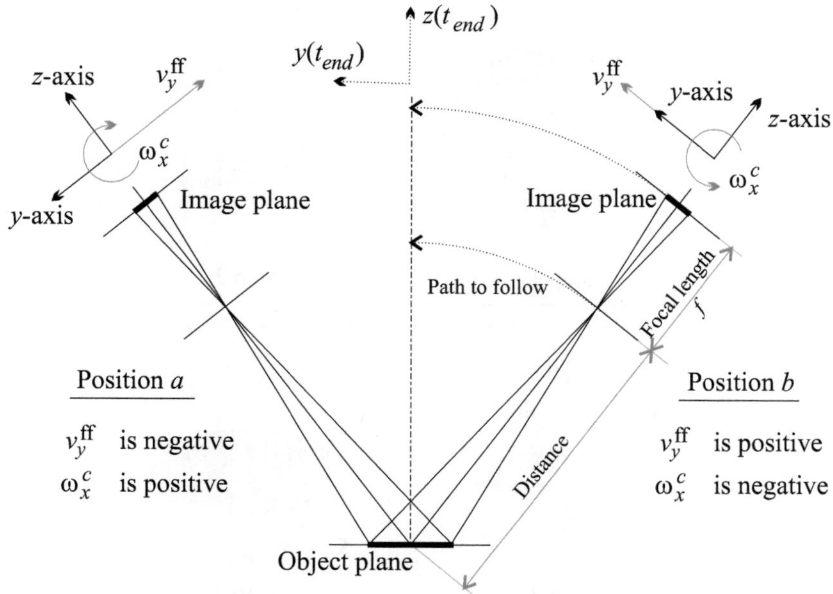

Fig. 5.3. Ambiguity in the determination of the correct sense of step 3 in Fig. 5.1: the 2 drawn camera positions give almost identical images when using a circle as alignment object

The verification of a valid global end condition is also a very sensitive operation. Since several motions occur simultaneously and since these motions, based on the changed visual information, are not independent from each other, small oscillations and/or drift in the robot motion emerge. Then, all the end conditions for the separate controlled motions will never be fulfilled simultaneously. Hence the global end condition will not be reached.

5.1.2 3D Alignment Using an Ellipse

Besides the practical problems, a *circle* offers, as mentioned above, the advantage that the real diameter of the circle is always related to the longest axis of the perceived ellipse, independent from the orientation of the image plane w.r.t. the object plane. A circle does however not give any information for the rotation around the optical axis. The use of an *ellipse* as the alignment object solves this problem. Aligning the longest axis of the ellipse with the x-axis of the task frame, defines the global desired end position up to 180°. Figure 5.4 shows the four steps in the 3D alignment with an ellipse.

The use of an ellipse however creates a new problem in step 3 of Fig. 5.4, which tries to make the image plane parallel to the object plane. Here, not only the perceived height of the ellipse (being the shortest axis) but also the perceived width (being the longest axis) needs to be maximized. Hence, the

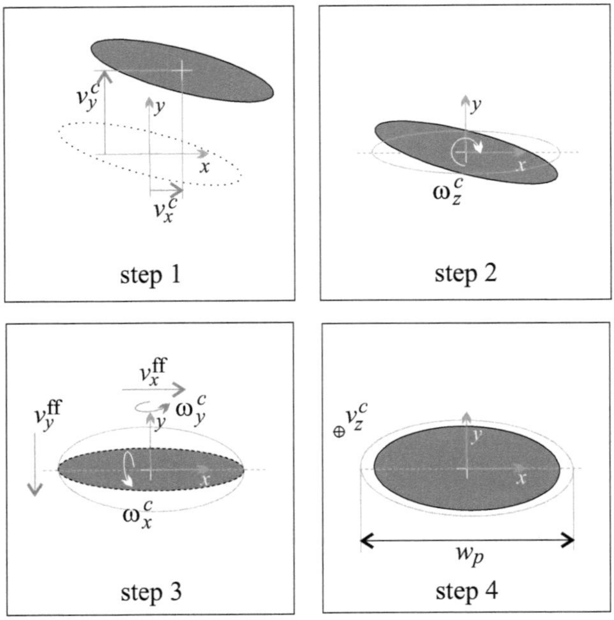

Fig. 5.4. Sequence of images in the four steps of the 3D alignment task using an ellipse as alignment object

problems for step 3 in the alignment with a circle, here arise twice. Moreover, in contrast to a circle, the accuracy of width and height measurements of the perceived ellipse, strongly depend on the accuracy at which step 2 is performed and maintained. A small rotation around the optical axis (z-axis) has a profound influence on the width and height measurements.

Hence, the practical implementation remains very difficult, is very noise sensitive and the resulting positioning is not satisfactory. The achievable accuracy with a stable control is very limited. Increasing the proportional control gains quickly results in restless and oscillatory behaviour. Also the simultaneous fulfillment of all the end conditions remains very difficult.

5.1.3 3D Alignment Using a Rectangle

The use of a rectangle as the alignment object solves most of the previously mentioned problems. In the desired end position of the camera, as shown in Fig. 5.5, the longest side of the rectangle lies horizontal, the shortest side lies vertical. This defines the global end position up to 180°. The rectangle alignment task is described as follows:

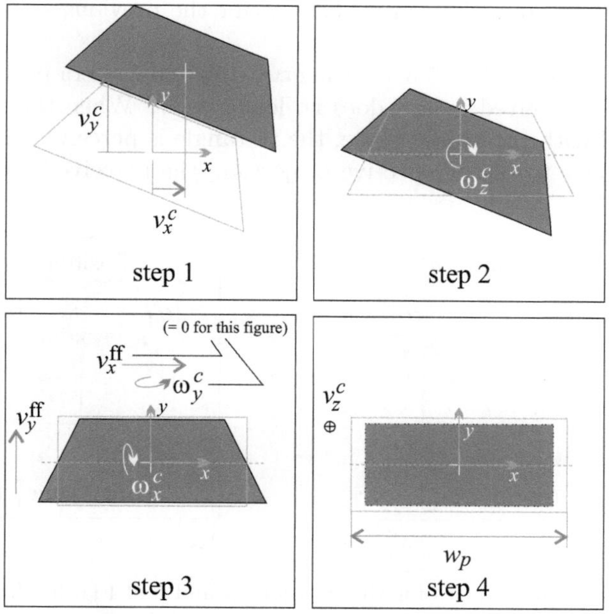

Fig. 5.5. Sequence of images in the four steps of the 3D alignment task using a rectangle as alignment object

Position the camera in such a way that (i) the optical axis passes through the center of the rectangle (using 2 DOF), that (ii) the longest side of the rectangle lies horizontal, i.e. parallel to the x-axis (using 1 DOF, defined up to 180°), that (iii) the image plane is parallel to the object plane (again using 2 DOF) and that (iv) the distance between the object plane and the image plane amounts to a given value (setting the last DOF).

Figure 5.5 shows the four steps in realizing the above alignment task. They are:

1. Center the perceived image by translating in x and y-direction.
2. Align the longest axis of the rectangle with the x-direction by rotating around the z-axis (i.e. the optical axis).
3. Position the image plane parallel to the object plane, by rotating around the x-axis and respectively the y-axis in order to make the perceived opposite sides of the rectangle parallel. Simultaneously translate along the y-direction and respectively the x-direction, so that the actually followed camera path is arc shaped with the center of this path at the object. The perceived image will slowly evolve from a trapezoid alike to a rectangle. The image plane is now parallel to the object plane.
4. Adjust the distance between the image plane and the object plane by a translation over the z-axis.

With each new step, the control actions for the previous steps are maintained.

The ambiguity of Fig. 5.3, showing two different camera positions resulting in the same perceived image, does no longer exist. When the image plane is not parallel with the object plane, the rectangle is perceived as a (distorted) trapezoid. Due to the perspective projection, the perceived opposite sides of

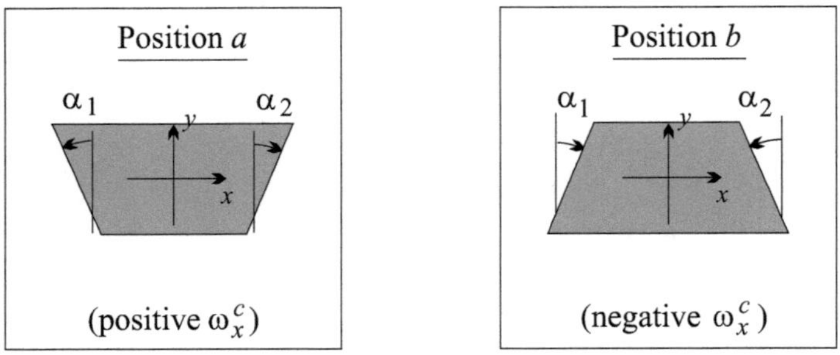

$$\omega_x^c = \sim \alpha_1 - \alpha_2 = \Delta\alpha$$

Fig. 5.6. Perceived images for the positions a and b of Fig. 5.3 for the definition of the x-alignment angle error $\Delta\alpha$: the sense of ω_x^c follows unambiguously from the angles α_1 and α_2.

the rectangle do not lie parallel. As shown in Fig. 5.6, for camera position a, the largest side of the trapezoid lies in the top half of the image. For camera position b, the largest side of the trapezoid lies in the bottom half of the image. Hence the sense of the action for step 3 is unambiguously defined. This is true for both the rotation around the x-axis (for which the figure is drawn) and the rotation around the y-axis.

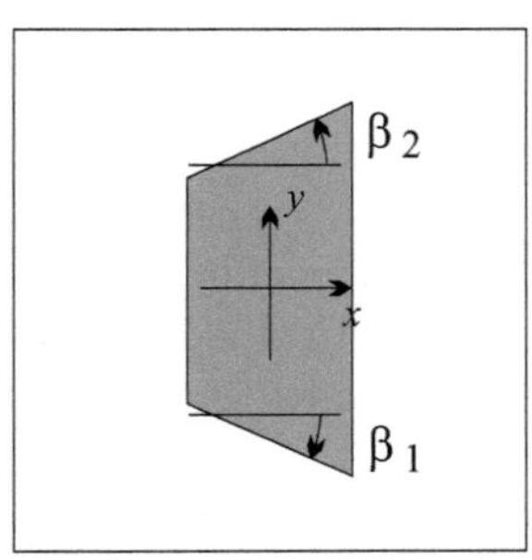

$$\omega_y^c = \sim \beta_1 - \beta_2 = \Delta\beta$$

Fig. 5.7. Definition of y-axis alignment angle error $\Delta\beta$

5.2 Used Algorithms

This section briefly reviews the used algorithms in the processing of the vision information.

The x- and y-position control use the image features XS_{ra} and YS_{ra} respectively, which are the x-signed and y-signed relative area parameters. XS_{ra} is equal to the number of object pixels with positive x-coordinates minus the number of object pixels with negative x-coordinates[5]. YS_{ra} is equal to the number of object pixels with positive y-coordinates minus the number of object pixels with negative y-coordinates. A detailed formulation of these parameters is given in Sect. A.5. The velocities v_x^c and v_y^c, shown in step 1 of Figs. 5.1, 5.4 and 5.5, command the camera frame to a position where the image feature parameters XS_{ra} and YS_{ra} are equal to zero.

The z-orientation control is based on the image feature XYS_{ra}, which gives the difference in object pixels lying on the one hand in the first and third quadrant and on the other hand in the second and fourth quadrant of the image as defined in Sect. A.5. The velocity ω_z^c, shown in step 2 of Figs. 5.1, 5.4 and 5.5, commands the camera frame to an orientation where the image feature XYS_{ra} equals zero. If $XYS_{ra} = 0$, then the object lies centered and horizontally in the image.

The rotations about the x- and y-axes of step 3 need to maximize the height in the case of a *circle* and the height and the width of the alignment object in the case of an *ellipse*. In contrast to the measurement of the image feature parameters XS_{ra}, YS_{ra} and XYS_{ra}, the measurement of width or height (pix) (of circle or ellipse) is not based on an averaging principle and has therefore no sub-pixel accuracy.

If a *rectangle* is used as alignment object, then step 3 is based on Fig. 5.6. The rotation ω_x^c about the x-axis[6] follows from the difference in the measured angles α_1 and α_2, defined in Fig. 5.6

$$\omega_x^c = K_{\theta_x} \Delta\alpha = K_{\theta_x}(\alpha_1 - \alpha_2). \tag{5.1}$$

Equally, the rotation ω_y^c about the y-axis follows from the difference in the measured angles β_1 and β_2, defined as the angles of bottom and top edges as shown in Fig. 5.7,

$$\omega_y^c = K_{\theta_y} \Delta\beta = K_{\theta_y}(\beta_1 - \beta_2). \tag{5.2}$$

The angles α_1, α_2, β_1 and β_2 and the width w_p of the rectangle are measured using the relative area contour detection described in Sect. 3.6. To this end, small windows lying centered at left and right edges on the one hand, and at bottom and top edges on the other hand, are used.

[5] w.r.t. the camera frame

[6] The actual used control gains are easily derived from the figures that present the results.

The distance along the z-axis between the image plane and the object plane[7], ${}^{obj}z$, follows from the known real width of the alignment object x_w (mm) and the perspective view model of the camera given by (3.22) in Sect. 3.9. The commanded velocity v_z^c along the z-axis, shown in step 4 of Fig. 5.5, equals

$$v_z^c = -K_z(z^d - z^m) \tag{5.3}$$

with

$$z^m = ({}^{obj}_{cam}z^m) = -\frac{x_w\, f}{w_p\, \mu_p}. \tag{5.4}$$

5.3 Experimental Results

5.3.1 Experimental Setup

The experimental setup consists of a KUKA361 robot, with end effector mounted Sony CCD camera (type XC77bbce). The focal length of the camera lens is about 6 mm. The robot controller is implemented on a transputer board (type T801) equipped with analogue input and output cards to control the robot actuators and to measure the actual robot position. The grabbed image is a 512 × 512 matrix of grey intensity values ranging from 0 to 255 (8 bit). The image is processed on a digital signal processor (DSP) (type Texas Instruments C40). The communication between the DSP and the Transputer is carried out on a transputer link.

The algorithms, described in Sect. 5.2, are implemented in C and executed on the DSP on a sub-image of 256 × 256 pixels at a frequency of 12.5 Hz. Decreasing the used image size enables a frequency of 25 Hz, which is the non-interlaced video frequency. A frequency of 12.5 Hz, however, proves to be sufficient for a smooth and stable control. Due to the larger field of view, the lower sample frequency is preferred. At the start of each experiment the object has to lie, at least partly, in the camera's field of view.

5.3.2 Global Remarks

Figures 5.8 to 5.11 give some experimental results for the 3D alignment tasks using either a circle, an ellipse or a rectangle as alignment object. All results are shown as a function of time. The time instant at which the end condition for a given step is fulfilled and a new control action is taken, is indicated with the step number. All end conditions are specified as a maximum allowable remaining feature error. For example, step 1 'ends' when the feature parameters XS_{ra} and YS_{ra}, which correspond to the distances ${}^{obj}x$ and ${}^{obj}y$ between

[7] In contrast to all previous features, this feature is not truly image based.

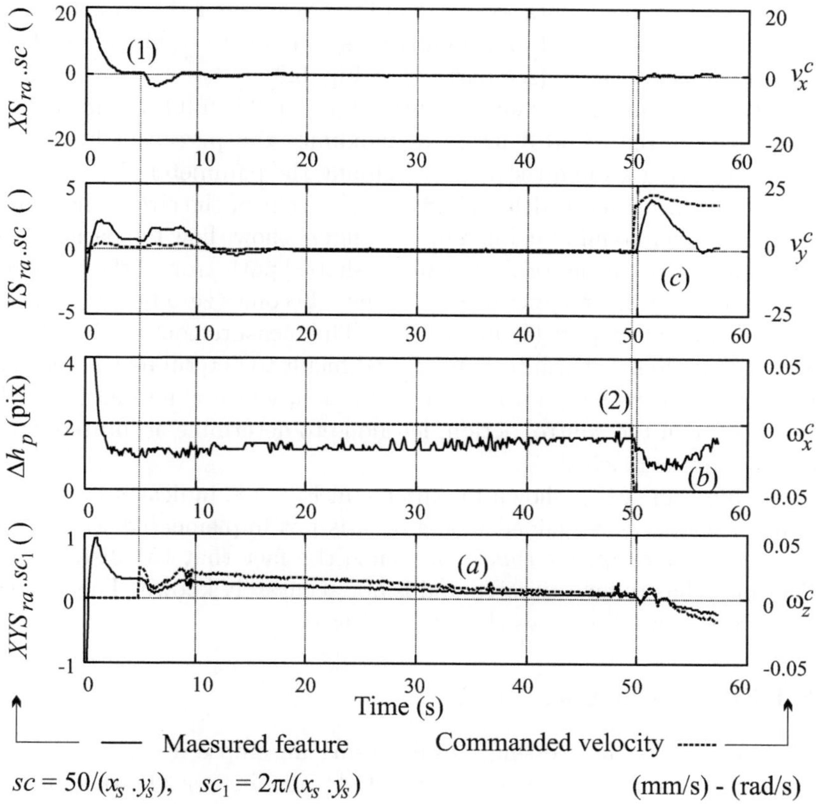

Fig. 5.8. Measured features and commanded velocities versus time while testing the fitness of a circle as alignment object

the optical axis and the center of the object (along x- and y-directions respectively), are smaller than a given threshold. Time periods in the figures are denoted by letters.

For a given TF direction, all figures show the measured image feature together with the commanded velocity, in order to better understand the physical motion. The goal of each control action is to minimize the corresponding feature value. Typically in image based visual servoing (in contrast to position based visual servoing), these feature errors are expressed in pixels or are dimensionless (in our case by dividing by the image size $x_s y_s$ as indicated by the scale factors sc and sc_1 in the figures).

5.3.3 Circle Alignment Task

Figure 5.8 shows the (test) results using a circle as alignment object. At time instant (1), the distance errors along x- and y-direction are smaller than the given threshold and the control for step 2 of Fig. 5.1 starts. An accurate measurement of the orientation error of the circle about the z-axis by the parameter XYS_{ra} is difficult and slightly noisy. To assure stability, a sufficiently slow control is necessary (see part (a) in Fig. 5.8).

At time instant (2), the end condition for step 2 is fulfilled and the control of step 3 is tested. Step 3 needs to maximize the perceived height of the circle by rotating around the x-axis. Defining the parameter Δh_p (pix) as the difference in measured width and measured height of the circle, the control of step 3 boils down to minimizing Δh_p. The figure shows how the parameter Δh_p changes when moving according to an arc shaped path around the center of the circle. The experiment shows that Δh_p never becomes zero but it does reach a minimum value (see part (b) in Fig. 5.8). The measurement of Δh_p, however, is very noisy. Hence, it will be extremely difficult to determine whether Δh_p is increasing or decreasing during a given motion, which is needed to check the correct sense of the control action, let alone to determine accurately whether the minimum is reached.

The jump in XY_{ra}, shown in part (c) in Fig. 5.8, indicates that the feed-forward control, as explained in Fig. 5.2, is not implemented with complete accuracy. The reason for this is of course the fact that the distance to the object plain has to be estimated and is not accurately known. The y-direction control, however, eliminates the occurred errors.

5.3.4 Ellipse Alignment Task

Figure 5.9 shows the (testing) results using an ellipse as alignment object. The y-orientation control is not executed. At time instant (1), the optical axis passes through the center of the ellipse. This starts the z-orientation control, which is very stable and more accurate, since the longest axis of an ellipse, in contrast to a circle, is clearly measurable. The z-orientation control fulfills its end condition at time step 2.

Now the fitness of a change in Δh_p, as control signal for the x-orientation control, is tested. In part (a) in Fig. 5.9 the camera is again commanded to move on an arc shaped path around the center of the ellipse. This time, the figure also gives the supposed control direction dir (=1 or =-1) by showing the signal $\Delta h_p dir$. This measurement behaves oscillatory and is unfit to be used as the basis for a control action. The determination of the correct sense of the needed action goes wrong due to the noisy measurements and the ill-conditioned alignment, as previously explained.

As part (b) in Fig. 5.9 shows, the x-orientation control still influences the y-position control, because the distance to the object plane is not accurately estimated.

5.3.5 Rectangle Alignment Task

The results for two experiments using a rectangle as alignment object are shown in Figs. 5.10 and 5.11.

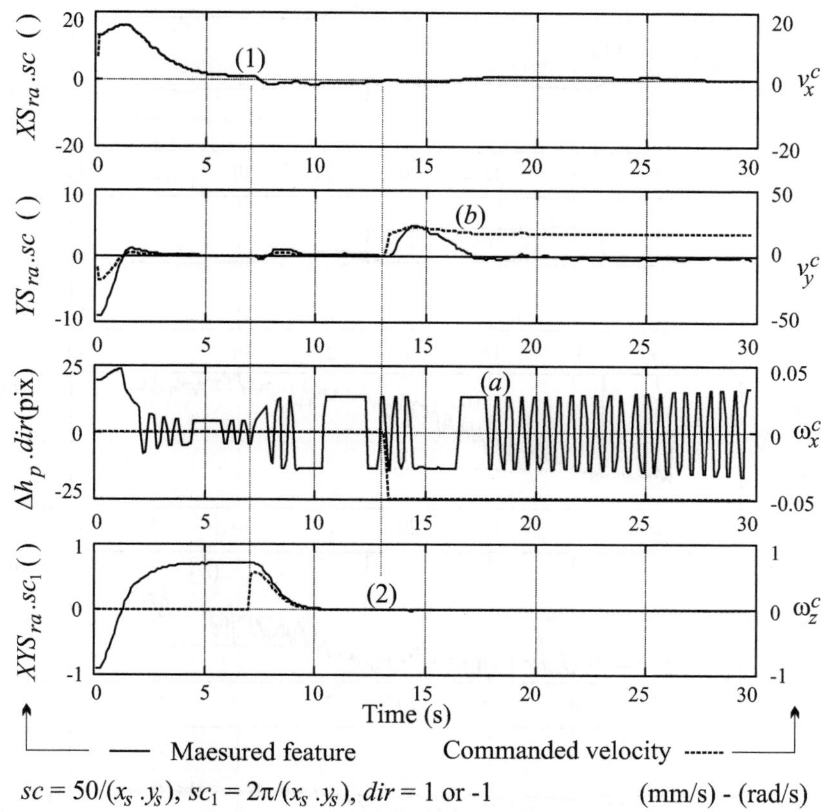

Fig. 5.9. Measured features and commanded velocities versus time while testing the fitness of an ellipse as alignment object

The first experiment tests the suitability of the $\Delta\alpha$ and $\Delta\beta$ measurements, according to the demand for parallel opposite sides explained in Figs. 5.6 and 5.7, as a basis for the x- and y-orientation control. At time instant (1) in Fig. 5.10, the first step is accomplished. The optical axis of the camera now passes through the center of the perceived 'rectangle'. The rectangle lies almost horizontal and the (weak) end condition of the z-orientation control is (almost immediately) fulfilled. During the period (a), the task frame rotates around the (imaginary) horizontal axis of the rectangle at a fixed velocity, in order to measure the change in $\Delta\alpha$. During the period (b), the task frame rotates around the (imaginary) vertical axis of the rectangle at a fixed velocity, in order to measure the change in $\Delta\beta$. The orientation measurements clearly contain the searched orientation information but are still fairly noisy. This necessitates filtering or averaging out of the measurements.

Figure 5.11 gives the final results in the rectangle alignment task. At time instant (1), step 1 of Fig. 5.5 is accomplished. The remaining x- and y-feature errors are smaller than the given threshold. This activates the z-orientation

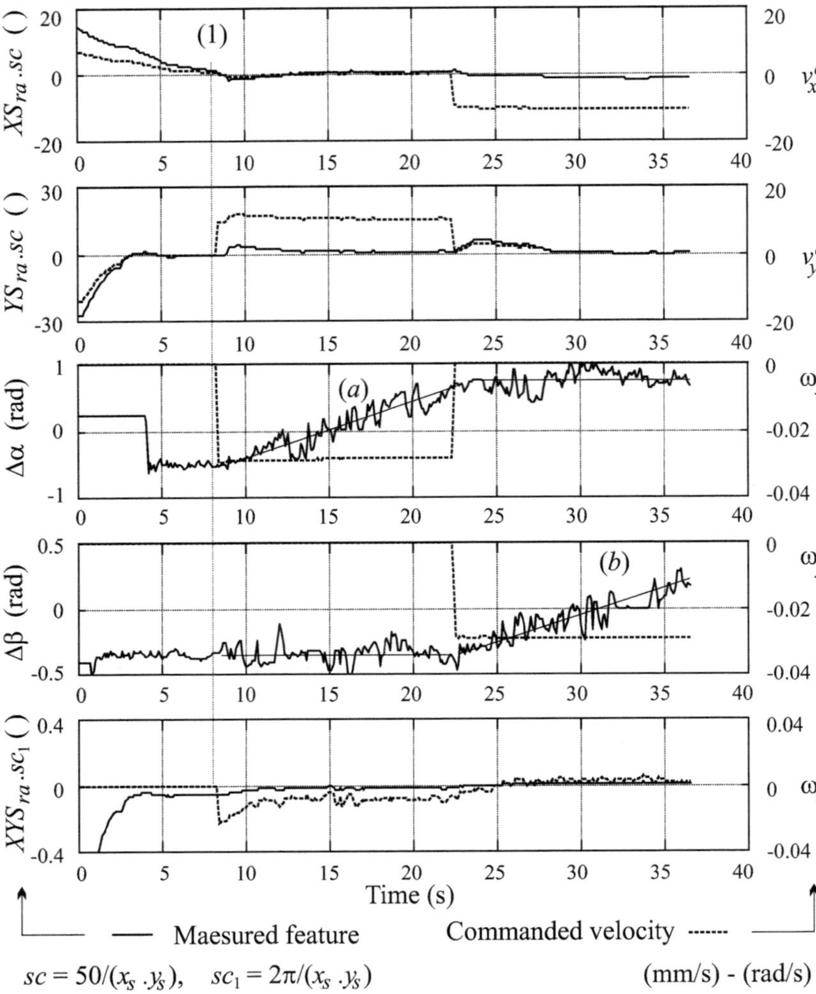

Fig. 5.10. Measured features and commanded velocities versus time while testing the fitness of a rectangle as alignment object

control, which reaches its goal at time instant (2), according to step 2 of Fig. 5.5. Also the x- and y-orientation control, activated separately in parts (a) and (b) respectively and both together in part(c), live up to the requirement. During this experiment, the orientation measurements are averaged out over 10 samples. At time instant (3) the end conditions for the first three steps (of Fig. 5.5) are simultaneously fulfilled. Now the distance adjustment can start.

To test the accuracy of the alignment task, all control actions are now turned off and the robot is commanded to translate along the z-axis of the camera frame until the desired distance to the object is reached. Finally, after about 35 seconds, the alignment task reaches its global goal.

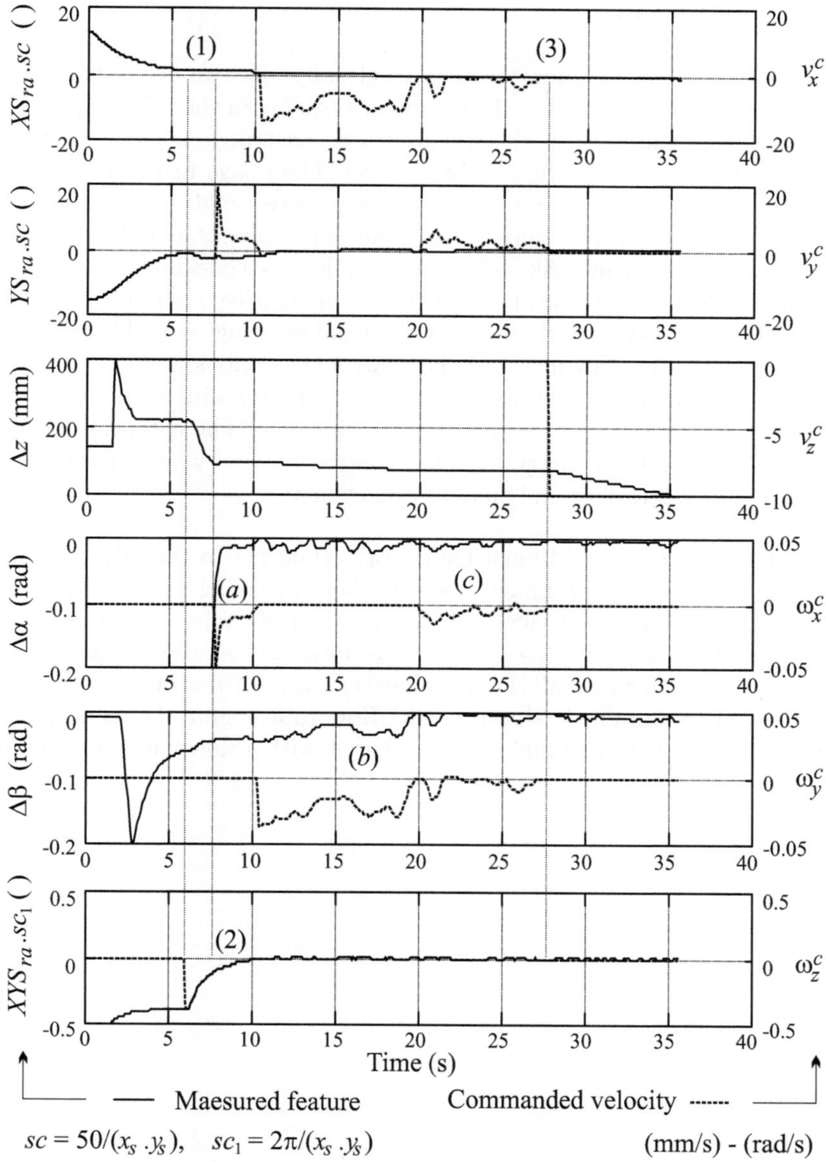

Fig. 5.11. Measured features and commanded velocities versus time for the 3D alignment of the eye-in-hand camera with a rectangle as alignment object

During the last step, the robot moved over a distance of 78 mm along the z-axis. The center of the object shifted by 0.75 pix. For the given setup this corresponds to an orientation error between the optical axis and the object normal of 0.3°. In view of the limited accuracy at which the camera frame is defined w.r.t. the end effector frame, this result is more than satisfactory.

5.4 Conclusion

This section presents a new strategy to visually align an eye-in-hand camera with a given object. For this 3D alignment task, i.e. in the relative positioning of the camera frame w.r.t. the given object, a rectangle is to be preferred as alignment object over a circle or an ellipse. The image features and likewise the corresponding control actions in the case of a rectangle as alignment object are easier to extract and more robust than in the case of a circle or an ellipse.

The 3D alignment task using a rectangle is successfully and accurately executed. Thanks to the averaging effect of the relative area image processing methods, the implemented image based visual servoing control is quite robust and always stable, provided that the control constants stay limited.

The 3D alignment task illustrates the visual servoing capabilities of the robot using image based control of the task frame. The extension to a complete 3D following system is, from the control point of view, straightforward, given a sufficiently slow and smooth object motion.

Some aspects are still open for investigation: (1) In the alignment tasks, the goal position of the image plane is chosen parallel to the object plane. From a physical point of view, this seems a logical choice. But a non-parallel alignment task may perhaps give better results. (2) Can the alignment task be executed more quickly if all the control actions are started simultaneously with guaranteed (maintained) stability? (3) How robust is a 3D visual following control based on the rectangle alignment task with a slowly moving rectangle?

6

Planar Contour Following of Continuous Curves

6.1 Introduction

In planar force controlled contour following, the robot is holding a tool while following the contour of a workpiece. When pose and shape of the workpiece are unknown, the force sensor is used to modify, or even generate, the tool trajectory on-line, which is referred to as *force tracking*, explained in Sect. 3.5. Due to the limited bandwidth of the sensor-based feedback control loop (loop d^{tr} in Fig. 3.1), the execution speed of the task is restricted in order to prevent loss of contact or excessive contact forces.

This chapter shows how the performance of a contour following task improves w.r.t. a pure force feedback controlled task by combining force control (tracking) and visual servoing. While maintaining the force controlled contact, the controller has to keep the camera, also mounted on the robot end effector, over the contour at all times. Then, from the on-line vision-based local model of the contour, appropriate feedforward control is calculated and added to the feedback control in order to reduce tracking errors.

The presented task exploits the variable EOL configuration, which is the most adequate to assure easy image processing and the most challenging in the sense of control.

The approach presented in this chapter can be applied to all actions that scan surfaces along planar paths with a *rotationally symmetric* tool: cleaning, polishing, or even deburring ... It is especially useful for one-off tasks in which accurate positioning or calibration of the workpiece is costly or impossible.

Section 6.2 reviews the motivation for the chosen EOL setup. Section 6.3 gives the full planar contour following task description. Section 6.4 describes the details of the control approach. It deals with the problem of keeping the contour in the camera field of view in addition to maintaining the force controlled contact. It further explains how the vision data are matched to the actual position, and how the feedforward signal is calculated. Section 6.5 describes the experimental setup and presents the experimental results. Finally, Sect. 6.6 concludes this chapter.

6.2 EOL Setup

The combined vision/force contour following task, presented in this chapter, uses a variable EOL configuration. It corresponds to Task 2b of Fig. 4.5-bottom described in Sect. 4.3.

Figure 6.1 gives the complete setup. The task frame is connected to the tool (tip). The camera is mounted on the end effector ahead of the force sensor with the optical axis normal to the plane. As previously mentioned, mounting the camera in this way, results in local images, normally free from occlusion, and in a controllable camera position. The image feature of interest is placed on the optical axis (center of the image) by visual servoing, which makes the feature measurement less sensitive to calibration errors or distortions in the camera/lens system.

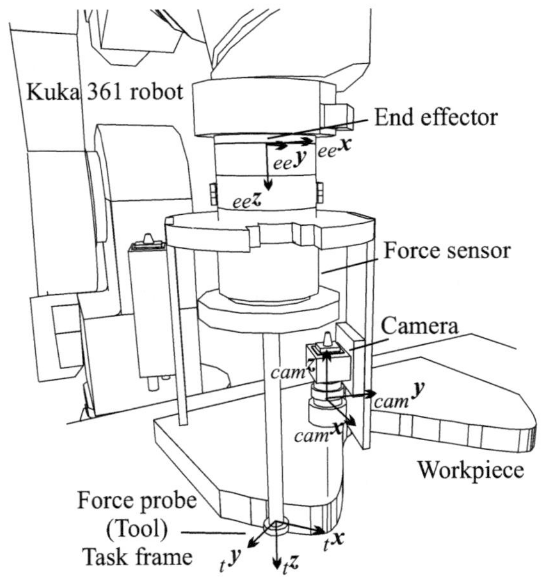

Fig. 6.1. Global overview of vision/force contour following setup

6.3 Task Specification

The task consists of following a planar contour with force feedback and vision based feedforward. The objectives are[1]:

1. move along the workpiece contour, i.e. along the $_tx$-axis, with given (tangential) velocity (direction type d^v),
2. maintain a constant normal (i.e. along the $_ty$-axis) contact force between tool and workpiece (direction type d^f),
3. keep the contour in the camera field of view by rotating the end effector (direction type d^{vs}),
4. keep the task frame tangent to the contour, i.e. keep the $_tx$-axis normal and the $_ty$-axis tangent to the contour, by rotating around the $_tz$-axis (direction type d^{tr}) and
5. generate the correct feedforward to be added to the rotation of the task frame in the plane in order to reduce tracking errors (direction type $d^{+\mathrm{ff}}$).

Fig. 6.2. Control actions for the contour following task with indicated control types according to the hybrid control scheme of Fig. 3.1

Figure 6.2 visualizes these desired actions. They are specified in the high level task description of our experimental control environment COMRADE [83].

[1] The respective direction types according to the control scheme of Fig. 3.1 are indicated between brackets.

The syntax and units used in this specification clearly indicate the control type. The following is an example of the program for the above task, expressed in the task frame indicated in Fig. 6.1:

TASK EXAMPLE: Planar contour following {
with task frame: variable - visual servoing
  x: velocity 25 mm/sec
  y: force 30 N
  z: velocity 0 mm/sec
  θ_x: velocity 0 rad/sec
  θ_y: velocity 0 rad/sec
  θ_z: track on velocities with feedforward
until relative distance > 550 mm}

The tracking action (objective 4 in the above list) has to reduce or eliminate the tracking error. This tracking error, as explained in Sect. 3.5, always depends in some way on a force measurement. The vision feedforward (action 5 in the above list) tries to avoid the occurrence of tracking errors in the first place. As shown in Sect. 3.2 by (3.8) the feedforward on a rotation is curvature based. Both (feedforward and tracking) actions control the same direction simultaneously. Hence, this gives the highest level of sensor integration: fusing force and vision information through *shared control* (type p3: $d^{tr+\mathrm{ff}}$).

6.4 Detailed Control Approach

This section describes the details of the control approach, explains the matching of the visually measured contour with the contact point at hand and implements the feedforward control.

6.4.1 Double Contact Control

Keeping the contour in the camera field of view while maintaining a force controlled contact corresponds to keeping a "virtual double contact". The first contact, point A in Fig. 6.3, is between workpiece and force probe. This contact point must move with the specified (tangential) velocity. The second (virtual) contact, point B in Fig. 6.3, coincides with the intersection of the optical axis and the contour plane.

To maintain this double contact, following control actions are taken[2]:

(1 - d^v) Velocity control: Move the task frame (point A) with the desired tangential velocity:

$$ {}_t v_x^c = {}_t v_x^d . \tag{6.1} $$

[2] For a description of the used notation see table 3.1 or list of symbols.

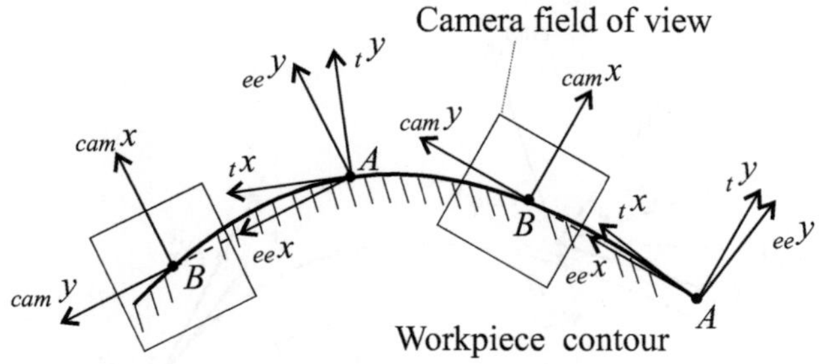

Fig. 6.3. Top view of contour for two time instants

(2 - d^f) Force control: Compensate the normal force error by applying a velocity in the $_ty$-direction:

$$_tv_y^c = K_y^f \, {}^tk_y^{inv}[\,_tF_y^m -_t F_y^d]. \tag{6.2}$$

K_y^f is the y-direction force control gain, ${}^tk_y^{inv}$ is the y-direction compliance of the tool and $_tF_y^m$ and $_tF_y^d$ are the measured and desired forces (from object to robot) in the y-direction of the task frame.
(3 - d^{vs}) Visual servoing: Rotate the end effector[3] around the $_{ee}z$-direction by:

$$_{ee}\omega_z^c = \omega_1 + \omega_2, \tag{6.3}$$

with

$$\omega_1 = K_{\theta z}^{vs} \; {}_{cam}^{\;c}x^m / r_{AB}. \tag{6.4}$$

Component ω_1 controls point B towards the contour. r_{AB} (mm) is the fixed distance between points A and B. Component ω_2 moves B tangent to the contour in B with velocity $\vec{v}_B$ (direction known, magnitude unknown) while compensating for the velocity of point A. This tangent based component is in fact a feedforward signal on the position control of point B as explained in Sect. 3.2. Its value follows from:

$$\vec{\omega_2} \times \vec{r_{AB}} + \vec{v_A} = \vec{v_B}. \tag{6.5}$$

According to the notations of Fig. 6.4 and neglecting the velocity $_tv_y^c$, ω_2 is solved as:

[3] From a practical point of view, the task frame is related to the end effector and not to some absolute world frame. Hence, moving or rotating the end effector frame will also move or rotate the task frame. To make the orientation of the task frame $_{abs}{}^t\theta_z$ independent from the rotation of the end effector by $_{ee}\omega_z^c$ (the visual servoing action), we have to redefine the relation between task and end effector by $-_{ee}\omega_z^c$ to compensate.

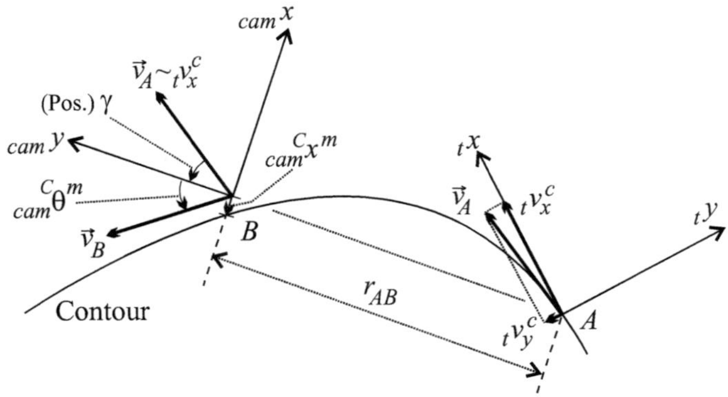

Fig. 6.4. Double contact with definition of variables

$$\omega_2 \cong -\frac{{}_t v_x^c \, \sin({}^{\,c}_{cam}\theta^m + \gamma)}{r_{AB} \, \cos({}^{\,c}_{cam}\theta^m)}, \tag{6.6}$$

with γ the angle between $\vec{v}_A$ and $_{cam}y$-axis.
(4 - d^{tr}) Force tracking: Compensate the tracking angle error $\Delta\theta_z \cong {}_t v_y^c / {}_t v_x^c$ by rotating the task frame w.r.t. the end effector frame.

This action does not change the actual position of the end effector. In theory it could take place in one control time step, since there is no mass displacement involved. From a practical point of view, however, the noise, which affects the identification of the tracking angle $\Delta\theta_z$, enforces the need of a low pass filter[4] resulting in:

$${}_t\omega_z^c = K_{\theta z}^{tr} \Delta\theta_z, \tag{6.7}$$

with $K_{\theta z}^{tr}$ the proportional control gain for the tracking direction (1/s).

6.4.2 Matching the Vision Data to the Task Frame

The second step, after controlling both camera and force probe positions, determines the link between the actual position of the task frame and the data of the contour, as collected by the vision system. This subsection describes how the contour data are measured[5], corrected and logged and how the logged contour data are then matched to the actual tool position.

6.4.3 Contour Measurement

The image processing algorithm determines (the absolute position of) the desired tool path, in several steps. Figure 6.5 gives an overview of these steps. The following explains these steps one by one. First, the Infinite Symmetric

[4] Also a sudden orientation change in the forward direction is not desirable.

[5] This measurement is unaffected by the camera/lens distortion, since the visual servoing control (6.3)–(6.6) keeps the optical axis of the camera close to the contour at all times.

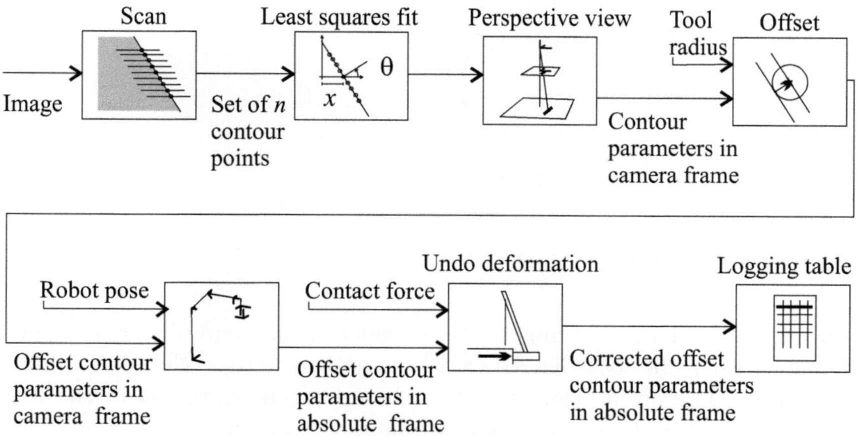

Fig. 6.5. Vision processing flow

Exponential Filter (ISEF), proposed by [73] and reviewed in Sect. 3.7, extracts local contour points (x_p, y_p) in the image space (pixels). Once a starting point on the contour is found, a narrowed search window is applied (which tracks the contour) to make the scanning of the subsequent contour points more robust and faster. In total n (e.g. $n = 9$) contour points are extracted lying symmetrically around the center (horizontal line) of the image, giving the data set:

$$[X_p Y_p] \text{ with } X_p = [x_p^1 \dots x_p^n]' \text{ and } Y_p = [y_p^1 \dots y_p^n]'. \tag{6.8}$$

The Total Least Squares (TLS) solution[6] of

$$\begin{pmatrix} & 1 \\ Y_p & \vdots \\ & 1 \end{pmatrix} \begin{pmatrix} \tan(-{}^c\theta) \\ {}^c x_p \end{pmatrix} = X_p \tag{6.9}$$

then determines the position ${}^c x_p$ (pixels) and orientation ${}^c\theta$ (rad) of the contour in the image. Figure 6.6 gives an example of these contour parameters.

Next, the position ${}^c_{cam}x$ (mm) of the contour w.r.t. the center of the camera frame is calculated according to the perspective view pin-hole model, given in Sect. 3.9,

$${}^c_{cam}x = -{}^c x_p \, \mu_p \, {}^c_{cam}z / f \tag{6.10}$$

[6] From a practical point of view, the available TLS algorithm is used to solve (6.9). Note however that a LS solution, which does not suppose errors on the column of 1's (or on the elements of Y_p) in (6.9), may give a slightly better and more justified result.

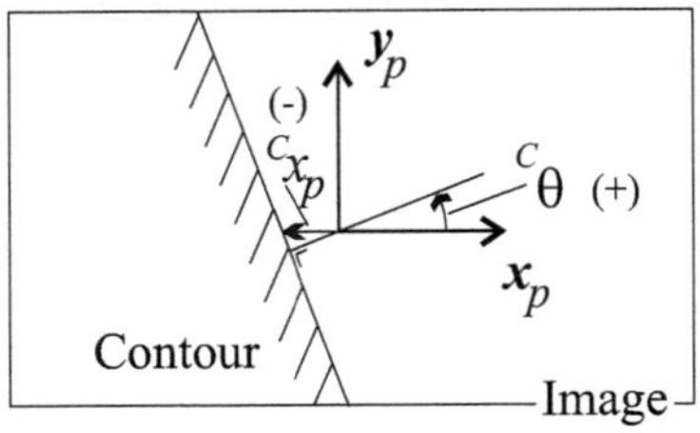

Fig. 6.6. Position and orientation of the contour in the image

with ${}^{C}_{cam}z$ the distance (mm) between camera frame and object plane (which is negative), f the focal length (mm) of the lens and μ_p the pixel dimension (mm/pix). The orientation does not change when expressed in the camera frame so ${}^{C}_{cam}\theta = {}^{C}\theta$. The contour parameters ${}^{C}_{cam}x$ and ${}^{C}_{cam}\theta$ are used in the visual servoing actions of (6.4) and (6.6).

The contour is now represented by the frame ${}^{C}_{cam}H$, with origin at the contour, x-axis normal and y-axis tangent to the contour:

$$
{}^{C}_{cam}H = \begin{pmatrix} \cos({}^{C}_{cam}\theta) & -\sin({}^{C}_{cam}\theta) & 0 & {}^{C}_{cam}x \\ \sin({}^{C}_{cam}\theta) & \cos({}^{C}_{cam}\theta) & 0 & 0 \\ 0 & 0 & 1 & {}^{C}_{cam}z \\ 0 & 0 & 0 & 1 \end{pmatrix}. \tag{6.11}
$$

This position is offset, normal to the contour, by the tool radius r_t (12.5 mm), to give ${}^{OC}_{cam}H$: the position of (one point on) the Offset Contour (still expressed in the camera frame),

$$
{}^{OC}_{cam}H = \begin{pmatrix} 1 & 0 & 0 & cos({}^{C}_{cam}\theta)r_t \\ 0 & 1 & 0 & sin({}^{C}_{cam}\theta)r_t \\ 0 & 0 & 1 & 0 \\ 0 & 0 & 0 & 1 \end{pmatrix} {}^{C}_{cam}H. \tag{6.12}
$$

Then, the relative frame position is transformed to absolute coordinates. This is based on the transformation from camera to end effector frame ${}^{cam}_{ee}H$, which is determined by calibration, and the forward kinematics of the robot[7] ${}^{ee}_{abs}FWK(q)$ (from end effector to absolute coordinates as a function of the robot joint coordinates q):

$$
{}^{OC}_{abs}H = {}^{ee}_{abs}FWK(q)\, {}^{cam}_{ee}H\, {}^{OC}_{cam}H. \tag{6.13}
$$

Finally, this single measurement is corrected to account for the deformation of the tool and the robot under the current contact forces. This correction is based on a linear spring model with stiffness ${}^{cam}k$. It shifts the contour position to the Corrected Offset Contour ${}^{COC}_{abs}H$ according to

[7] See appendix Sect. A.3.

$$ {}^{coc}_{abs}H = \begin{pmatrix} 1\,0\,0 & sin({}^{t}_{abs}\theta_z)_tF_y/{}^{cam}k \\ 0\,1\,0 & -cos({}^{t}_{abs}\theta_z)_tF_y/{}^{cam}k \\ 0\,0\,1 & 0 \\ 0\,0\,0 & 1 \end{pmatrix} {}^{oc}_{abs}H \qquad (6.14) $$

with ${}_tF_y$ the normal force and ${}^{t}_{abs}\theta_z$ the (orientation of the) tangent in the (current) contact point.

From each image, one measurement of the contour (represented by the frame ${}^{coc}_{abs}H$) is taken and logged. Together, these measurements represent the desired path to be followed by the tool center.

Note that the order in which the above calculations are performed, may change. For example, the tool offset computation (6.12) may also be executed as the last step just before the data logging.

6.4.4 Tool Position Measurement

The measured tool pose (${}^{t}_{abs}H^m$) follows straightforwardly from the robot pose and the relative pose of the tool w.r.t. the end effector, ${}^{t}_{ee}H$.

$$ {}^{t}_{abs}H^m = {}^{ee}_{abs}FWK(q).\,{}^{t}_{ee}H. \qquad (6.15) $$

The measured tool pose, however, does not take into account the tool (and robot wrist) deformation under the current contact force ${}_tF^m$. Due to the tool compliance ${}^tk^{inv}$, the actual tool pose, ${}^{t}_{abs}H^a$, differs from the measured pose. The measured pose thus has to be corrected in the sense of the acting contact force, this is normal to the contour (in the ${}_ty$-direction):

$$ {}^{t}_{abs}H^a = \begin{pmatrix} 1\,0\,0 & sin({}^{t}_{abs}\theta_z)\,{}_tF_y\,{}^tk^{inv} \\ 0\,1\,0 & -cos({}^{t}_{abs}\theta_z)\,{}_tF_y\,{}^tk^{inv} \\ 0\,0\,1 & 0 \\ 0\,0\,0 & 1 \end{pmatrix} {}^{t}_{abs}H. \qquad (6.16) $$

Equation (6.16) is similar to (6.14). They use the same contact force ${}_tF_y$ for the pose corrections (be it for a different point on the contour). Both equations use, however, different stiffness or compliance values. The tool and camera-setup compliances[8] are not equal!

6.4.5 Matching

The matching is based on corresponding absolute position of on the one hand the tool pose and on the other hand the visually measured path. The path ahead is logged in a look-up table, see Fig. 6.5, containing one entry of the absolute corrected contour pose (represented by ${}^{coc}_{abs}x$, ${}^{coc}_{abs}y$ and ${}^{coc}_{abs}\theta_z$ in correspondence to the result of (6.14)) for each image. The table is extended with

[8] The tool compliance and the camera-setup compliance are determined experimentally.

the arc length s, which will be used in the curvature calculation in the next subsection.

In order to eliminate the effects of the vision delay time T^{vs}, we need to compute the control parameters for the next time instant. Hence, not the current tool pose, but the predicted tool pose for the next time instant $(k+1)$, ${}_{abs}^{t}H^{a}((k+1)T^{vs})$, is used as a (interpolating) pointer into the look-up table:

$$
{}_{abs}^{t}H^{a}((k+1)T^{vs}) = \begin{pmatrix} 1 & 0 & 0 & cos({}_{abs}^{t}\theta_z)\,{}_{t}v_x^c\,T^{vs} \\ 0 & 1 & 0 & sin({}_{abs}^{t}\theta_z)\,{}_{t}v_x^c\,T^{vs} \\ 0 & 0 & 1 & 0 \\ 0 & 0 & 0 & 1 \end{pmatrix} {}_{abs}^{t}H^{a}(kT^{vs}). \tag{6.17}
$$

Figure 6.7 summarizes all the matching steps. The logged position which lies the closest to the predicted tool position is indicated as the matched position h.

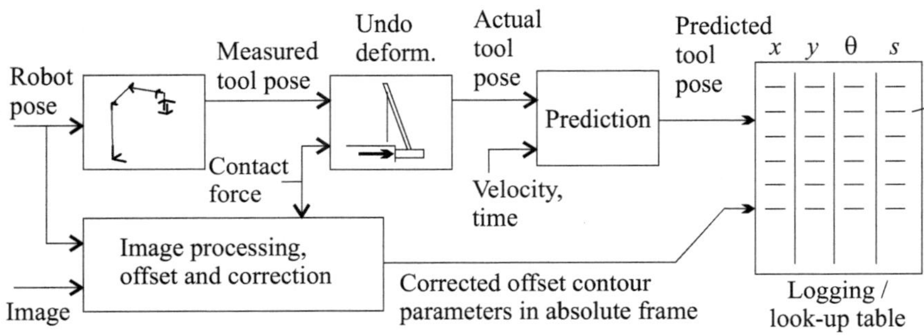

Fig. 6.7. Matching vision data to the tool frame

The advantage of the used position based matching is its simplicity. The position based matching will however fail if the contour itself moves. In this case, an arc length based matching method may give a solution. This option is not further investigated.

6.4.6 Calculating the Feedforward Signal

The implemented feedforward (loop $d^{+\text{ff}}$ in Fig. 3.1) has to avoid the occurrence of a tracking angle error $\Delta\theta_z$ using the following feedforward control signal:

$$
{}_{t}\omega_z^{\text{ff}} = -\kappa\,{}_{t}v_x^c \tag{6.18}
$$

with κ the curvature of the contour in the matched position and ${}_{t}v_x^c$ the tangent velocity of the tool, as explained in Sect. 3.2[9].

[9] Note the added minus sign in (6.18) w.r.t. (3.8). Hence, the correct sense of the feedforward signal depends on the setup!

An obvious way to calculate κ is the use of a fitted contour model. As previously explained in Sect. 3.8, however, this poses some problems: for a start, the calculation of the curvature from the fitted contour is very noise sensitive due to the second derivative. Furthermore, the fitted contour may differ from the real contour, especially for simple models or may be computationally expensive for more complicated models. Finally, not the curvature in one point is needed but the mean curvature for the arc length travelled during one time interval (T^{vs}).

In order to avoid all of the mentioned problems, the curvature is calculated as the mean change in orientation of the contour over the travelled arc length ds (see also Sect. 3.8):

$$\kappa = \frac{d\,{}^{coc}_{abs}\theta_z}{ds}. \tag{6.19}$$

κ results from the Total Least Squares [85] solution of a set of m first order equations (see (3.21)) lying symmetrically around the matched position (i.e. position h in Fig. 6.7). Note that there is no phase lag involved in the computation of κ.

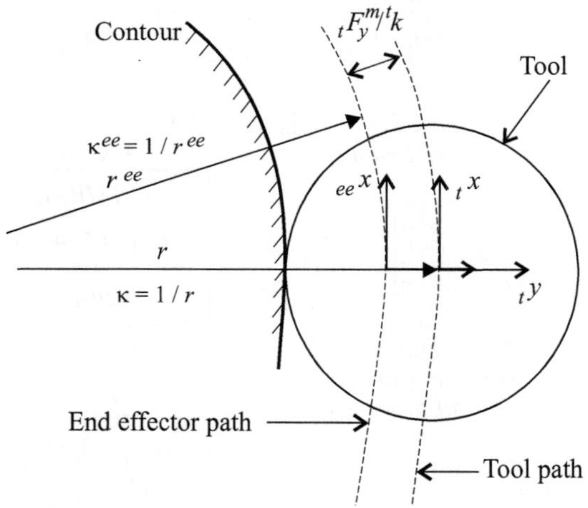

Fig. 6.8. Shift between end effector path and tool path due to the tool compliance

For the feedforward control too, the compliance of the tool needs to be taken into account. After all, due to the tool compliance, the z-axes of end effector and tool frames do not coincide. At an outer curve (e.g. curves 1 and 3 in Fig. 6.10), the end effector makes a sharper turn than the tool and thus needs to rotate faster. Figure 6.8 exemplifies this. The desired angular feedforward velocity ${}_{ee}\omega_z^{\rm ff}$ is therefore

$${}_{ee}\omega_z^{\rm ff} = -\kappa^{ee}\,{}_tv_x^{\rm c} \tag{6.20}$$

with

$$\kappa^{ee} = \frac{\kappa}{1 - \kappa \, {}_tF_y / {}^tk}, \tag{6.21}$$

${}_tF_y$ the normal contact force and ${}^tk$ the tool stiffness. The distance between end effector path and tool path amounts to ${}_tF_y/{}^tk$. This explains (6.21).

The feedforward velocity calculated according to (6.18)–(6.21), is added to the feedback control actions described in Sect. 6.4.1.

6.5 Experiments

First, the experimental setup is described briefly. Then the results are presented.

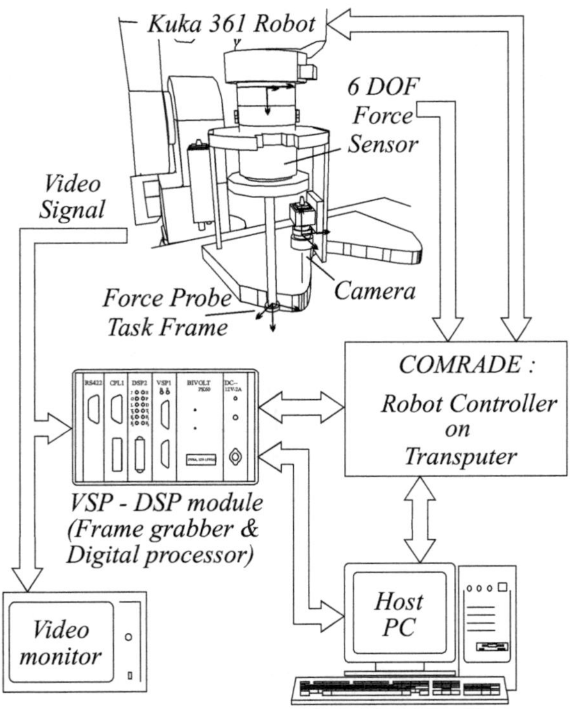

Fig. 6.9. Experimental setup

6.5.1 Experimental Setup

Figure 6.1 gives an overview of the setup. It consists of a KUKA 361 robot, with a SCHUNK force sensor together with an eye-in-hand SONY CCD XC77 camera with 6.15 mm lens. The CCD camera consists of 756 by 581 square pixels with 0.011 mm pixel width, from which however only a sub-image of

128 by 128 pixels is used. The dimensions of the camera mounting are described in appendix Sect. E.3.

Instead of the commercial controller, our own software environment COMRADE [29,83] is used. The robot controller and force acquisition are running at 100 Hz on a T801 transputer board. The image processing, the calculation of the camera position control signals, the matching and logging and the feedforward calculation are implemented on a TI-C40 DSP unit, with frame grabber and transputer link. The image processing and calculations are limited by the non interlaced frame rate of 25 Hz. Since the robot controller and image acquisition rates are different, the robot controller uses the most recent DSP calculations 4 times in a row.

The force probe is about 600 mm long with compliance $k_y^{inv} = 0.09$ mm/N. The distance r_{AB} is 55 mm. The camera is placed about 200 mm above the workpiece resulting in a camera resolution of about 3 pix/mm. The testing object has a sinusoidal form and is described in appendix Sect. E.1.

6.5.2 Results

Figure 6.10-top shows the uncorrected paths travelled by camera and task frame. The plotted task and end effector frame directions illustrate the variable relation between them during task execution. Figure 6.10-bottom shows the logged path (as measured by the vision system, offset by the tool radius and corrected for the camera-setup compliance) and the corrected tool path. The *maximum* error between these two is about 1 mm. This validates the matching method.

Figure 6.11-top compares the measured contact forces in the $_ty$-direction for two experiments with tangential velocity set to 25 mm/s: when *feedforward is used*, the desired normal contact force of 30 N is maintained very well ; *without feedforward*, both contact loss and contact forces of about 60 N occur.

The tangential velocity is limited to 25 m/s when only feedback is used, but can be increased to 75 mm/s without loss of contact for the given contour when applying vision based feedforward.

Figure 6.11-bottom shows the good correspondence between actually calculated and ideal feedforward signals for the given contour. The used method for the curvature calculation, however, levels out peaks in the curvature profile. This gives a less noisy feedforward signal but causes small remaining contact force errors at positions of high curvature.

6.6 Conclusion

This chapter presents a combined force tracking/visual servoing task in a variable EOL configuration. It shows how the quality of force controlled planar contour following improves significantly by adding vision based feedforward on

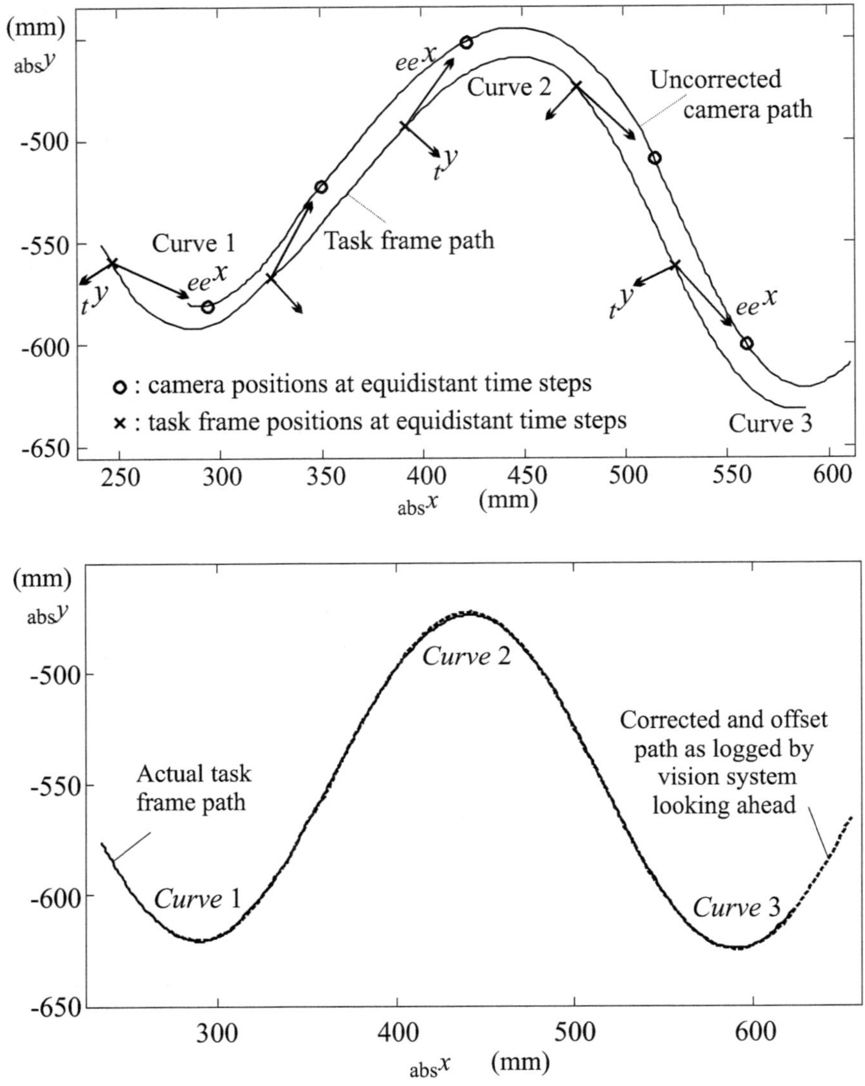

Fig. 6.10. Paths travelled by camera and task frame

the force tracking direction. This reduces tracking errors, resulting in a faster and more accurate execution of the task. The feedforward signal is calculated from the on-line generated local data of the contour. To get a good match between tool path and visually measured contour data, compliances of both tool and camera-setup are modelled and incorporated in the measurements.

Vision based feedforward is fused with feedback control in a tracking direction. As the tracking control itself is based on force measurements, this is an example of shared control.

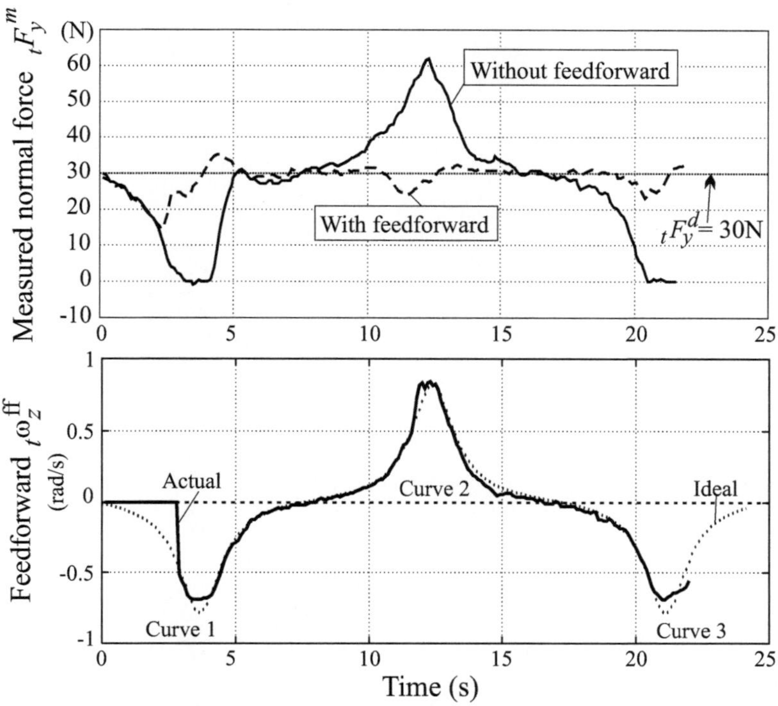

Fig. 6.11. Measured normal contact forces $_tF_y^m$ without and with use of vision based feedforward on the tracking direction $_t\theta_z$ (**top**); Actual and ideal feedforward signals for the given contour (**bottom**)

Keeping the contour in the camera field of view, while maintaining a force controlled contact, however, imposes additional requirements on the controller. This double control problem is solved using the redundancy for rotation in the plane, which exists for rotationally symmetric tools. The key in this solution is redefining the relation of the task frame with respect to the end effector in order to keep the task frame tangent to the contour, while rotating the end effector to position the camera above the contour. Hence, the orientations of task and end effector frames are controlled independently, hereby fully exploiting the variable EOL configuration.

The experimental results validate the used approach.

7

Planar Contour Following at Corners

7.1 Introduction

The solution presented in the previous chapter is not directly applicable to contours which contain places with extremely high curvature or corners. This chapter shows how the combined vision/force approach of the previous chapter can be adapted to also improve contour following tasks at corners[1].

The camera watches out for corners. If a corner is detected, a finite state control is activated to successfully take the corner. The corner detection and the finite state controller are the main new contributions w.r.t. the approach presented in Chap. 6.

The *global setup* of Fig. 6.1 remains. Only in this case, the contour contains a corner in the path ahead. As in the previous chapter, the camera is mounted on the end effector ahead of the force sensor with the optical axis normal to the object plane. The approach is still position based and endpoint open-loop. It uses a calibrated camera-setup.

Section 7.2 describes the corner detection. Section 7.3 treats the new control approach issues. The basic control scheme, shown in Fig. 3.1, is augmented to a three layer control structure with a finite state machine. Section 7.4 presents the experimental results. Finally, Sect. 7.5 concludes this chapter.

7.2 Path Measurement and Corner Detection

Section 6.4.2 and Fig. 6.5 already described the *contour measurement.* Figure 7.1 briefly reviews the results. It shows the measured contour, the actual contour (or corrected contour), the offset contour and the corrected offset contour. The latter is the desired tool (center) path. It further shows the measured tool center and the actual tool center. The latter must coincide with the corrected offset contour.

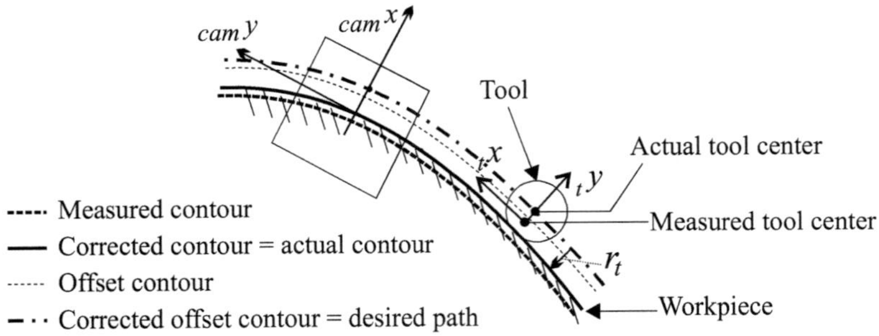

Fig. 7.1. Measured, actual, offset and corrected offset contours as measured or computed by the vision system on the one hand and measured tool and actual tool center path on the other hand

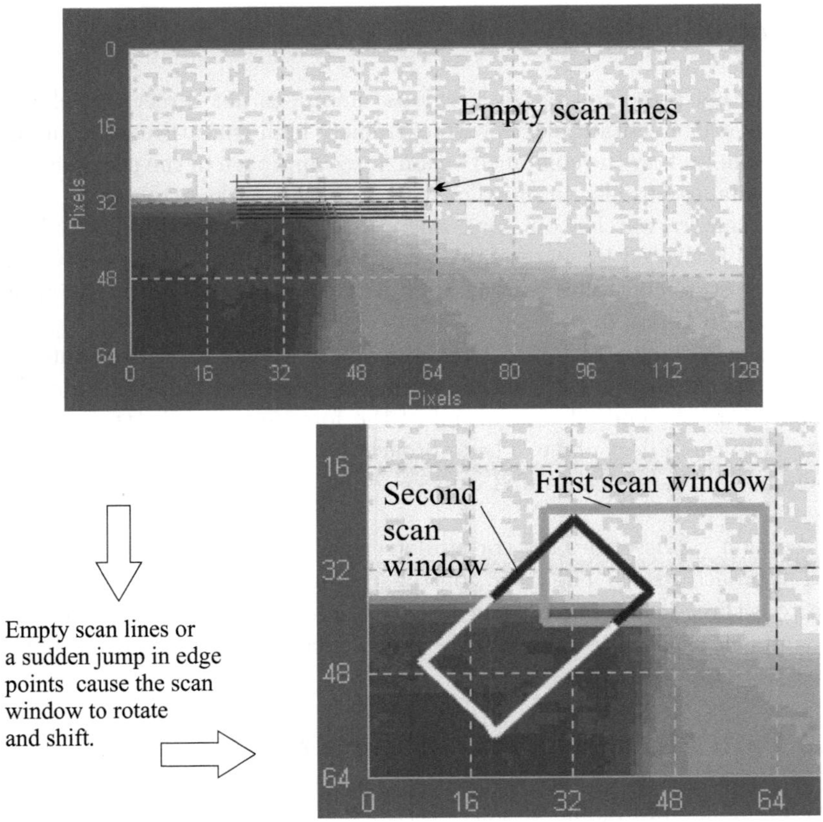

Fig. 7.2. Example of empty scan lines due to the presence of an edge and close-up of rotated and shifted scan window

The *corner detection* algorithm is based on the first step of the contour measurement, which extracts local contour points. If there is a sudden jump in coordinates of the extracted contour points $[X_p Y_p]$ or if there are empty scan lines due to the absence of an edge, the scan window is shifted and rotated[2], as shown in Figs. 7.2 and 7.3. The image is scanned again. If the second contour measurement is correct (no sudden jump nor empty scan lines) as shown in Fig. 7.3, its orientation is compared to the orientation of the previously logged contour. A corner is identified if the difference in orientation exceeds a given threshold. The position of the corner then easily follows from the intersection of two lines.

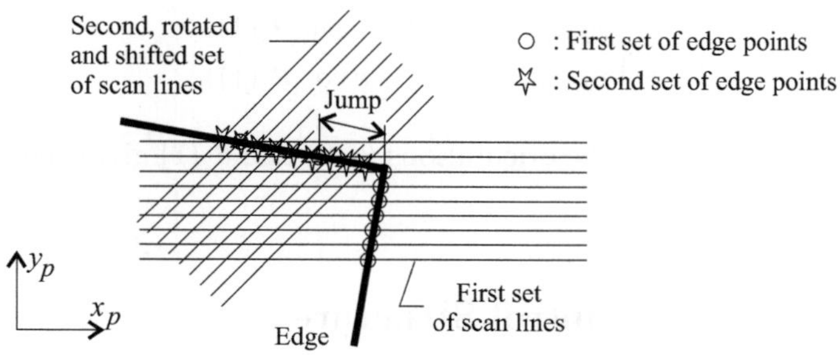

Fig. 7.3. Two consecutive sets of scan lines for corner detection

The exact location of the corner, however, is updated afterwards, when the optical axis of the vision systems is again positioned over the contour (as it was before the corner occurred) in which case the contour measurement is more accurate.

The corner position is offset by the tool radius. This offset corner is the starting point of the tool path around the corner, as shown in Fig. 7.6. We can calculate the offset corner as the intersection of the corrected offset contour, just before the corner occurred, with the corrected contour, after the corner occurred. These two lines are given by:

$$\begin{cases} x = {}^{coc}_{abs}\theta_z(1)\; y + {}^{coc}_{abs}x(1) \\ x = {}^{cc}_{abs}\theta_z(2)\; y + {}^{cc}_{abs}x(2) \end{cases} \tag{7.1}$$

with superscript cc indicating the corrected or actual contour. All parameters are expressed in absolute coordinates. For reasons of simplicity, the preceding sub- and superscripts are left away in the following. In order to get a stable, errorfree computation of the intersection of the two lines of (7.1), the equation is rewritten as

[1] No round but only sharp corners are considered.

[2] The rotation is 45°; the shift is large enough to avoid that the new scan window contains the expected corner.

$$\begin{cases} x(1) + d\cos(\theta_z(1) + \pi/2) = x(2) + l\cos(\theta_z(2) + \pi/2) \\ y(1) + d\sin(\theta_z(1) + \pi/2) = x(2) + l\sin(\theta_z(2) + \pi/2), \end{cases} \tag{7.2}$$

with (distance) parameters d and l. Only the solution for parameter d is needed[3]:

$$d = \frac{[x(1) - x(2)]\cos(\theta_z(2)) + [y(1) - y(2)]\sin(\theta_z(2))}{\sin(\theta_z(1) - \theta_z(2))}. \tag{7.3}$$

The resulting intersection of the two lines, which is the searched position of the offset corner, is then

$$\begin{cases} x_{corner} = x(1) - d\sin(\theta_z(1)) \\ y_{corner} = y(1) + d\cos(\theta_z(1)) \end{cases} \tag{7.4}$$

The difference between the orientations $\theta_z(2)$ and $\theta_z(1)$ gives the angle of the corner to be rounded.

7.3 Augmented Control Structure

Figure 7.4 illustrates the augmented control structure. The external Cartesian space control, using vision and force sensors around the low level servo controlled robot, corresponds to the control scheme of Fig. 3.1. The Cartesian control loop keeps the contour in the camera field of view, while maintaining a constant normal contact force as described in full detail in the previous chapter.

On top of the Cartesian space control, the sequencer and/or planner performs the overall control. It determines set-points, monitors transition conditions and adapts the control parameters for the different control states. The finite state machine, shown in Fig. 7.5 represents these different control states.

When a corner is detected, the tool slows down. At this point the end effector will have to turn very fast (without changing the direction in which the tool is moving) to keep the camera over the contour: an action that is limited by the maximum allowed velocity and/or acceleration of the robot. Here the robot dynamics play an important role. During this transition movement, the contour measurement is not very accurate and hence preferably not used. At sharp corners, moreover, the camera briefly loses the contour.

Once the tool (center) reaches the corner, the corner is taken at constant velocity while adapting the tangent direction (of the tool) by feedforward to follow the desired arc-shaped path. The desired angular feedforward velocity $_{ee}\omega^{\text{ff}}$ is

[3] Since the angles $\theta_z(1)$ and $\theta_z(2)$ are never equal, the denominator of (7.3) never becomes zero.

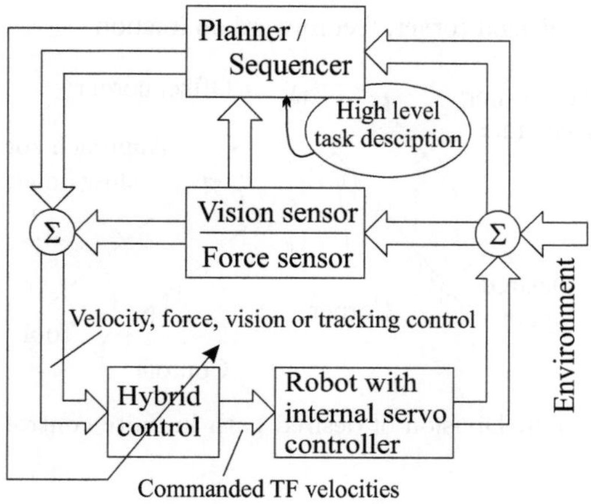

Fig. 7.4. Augmented three layer control structure

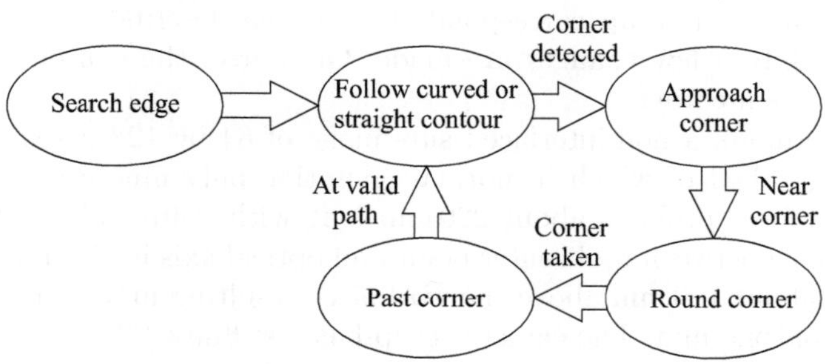

Fig. 7.5. Finite state scheme for planner

$$ {}_{ee}\omega^{\mathrm{ff}} = \frac{-{}_{ee}v_x}{r_t - {}_tF_y/{}^tk} \tag{7.5} $$

with ${}_{ee}v_x$ the tangent velocity of the end effector (or tool), r_t the tool radius, ${}_tF_y$ the normal contact force and ${}^tk$ the tool stiffness. Due to the compliance of the tool, the z-axes of end effector and tool (or task) frames do not coincide. At the corner the end effector makes a sharper turn than the tool (as previously illustrated in Fig. 6.8). The radius of this turn is smaller than the tool radius by a distance ${}_tF_y/{}^tk$. This explains (7.5).

After the corner is taken, the (tangent) velocity will gradually build up and the controller will return to the normal operation state.

Figure 7.6 gives an example which matches the desired path to the different control states. Some control specifications are given.

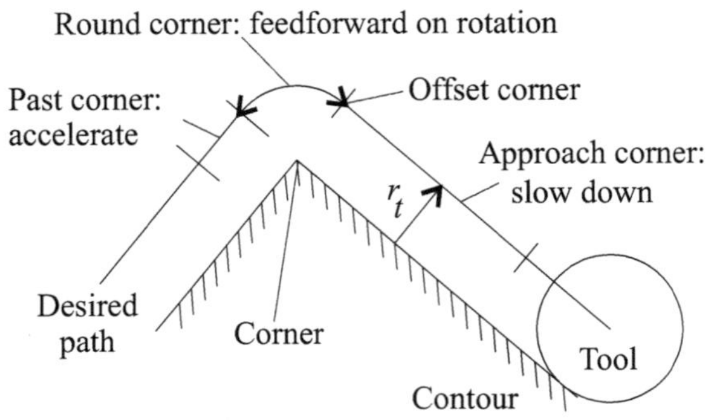

Fig. 7.6. Division of desired path in finite control states

7.4 Experimental Results

The experimental setup corresponds to the one described in Sect. 6.5 and Fig. 6.9. Only a few changes are made. Of course, the contour path ahead now contains a corner.

Furthermore, a non-interlaced sub-image of 64 by 128 pixels in 256 grey levels is grabbed of which in normal operation only nine centered lines are used. The force probe is about 220 mm long with stiffness ${}^{t}k = 13.2\,\mathrm{N/mm}$. The distance between tool center point and optical axis is 55 mm. The camera is placed about 100 mm above the workpiece resulting in a camera resolution of about 5.6 pix/mm. The camera-setup has a stiffness ${}^{cam}k = 21.2\,\mathrm{N/mm}$.

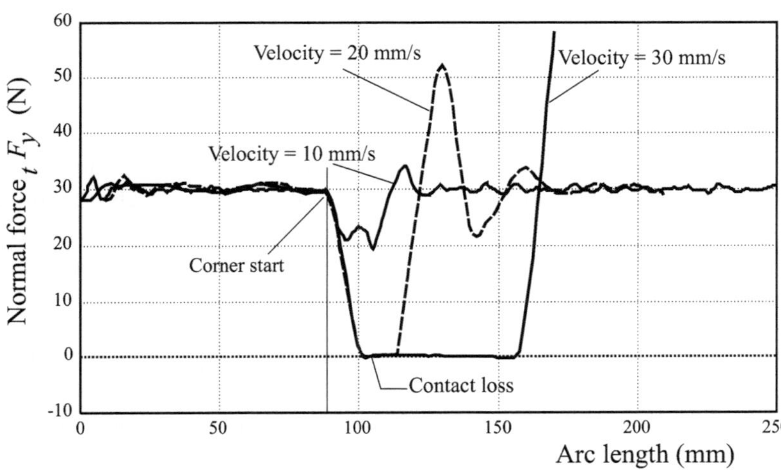

Fig. 7.7. Normal force versus arc length at different fixed velocities without using the vision system

The results of two experiments are presented. As a basis of comparison, the first experiment measures the contact forces while following the unknown contour[4] with a corner of 90°, without using any vision information. Figure 7.7 gives the results. The desired contact force is 30 N. In this experiment, contact is lost or excessive contact forces occur if the velocity is too high. To assure a fairly well maintained force profile, the maximum allowable tangent velocity is only 10 mm/s. However, experience shows that decreasing the tangent velocity further will not result in a better force profile: if the tangent velocity is too low, the tracking error identification (on velocities) will fail (see also Sect. 3.5) and the robot is unable to round the corner. The task literally gets stuck at the corner.

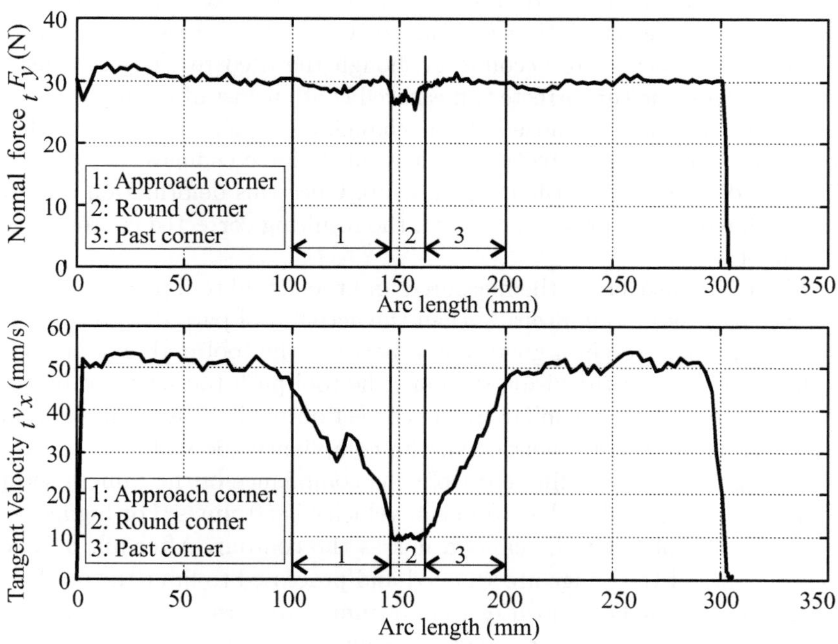

Fig. 7.8. Measured normal force (**top**) and tangent velocity (**bottom**) versus arc length for the contour following task using vision based feedforward at the corner

The second experiment implements the described finite state controller and corner detection in a combined vision/force contour following task of the same contour.

The top of Fig. 7.8 gives the measured normal force versus arc length for the combined vision/force control of the second experiment. The contour is successfully tracked without major changes in (normal) contact force. Only at the corner itself, the contact force is slightly smaller than desired, pro-

[4] The angle of the corner is of course also unknown.

bably due to imperfect corner measurement and/or non-modelled non-linear deformations of tool and robot wrist.

The bottom of Fig. 7.8 gives the measured tangent velocity versus arc length. At the corner the robot slows down to round the corner in the best conditions. After the corner the robot accelerates. For the second experiment, the normal operation velocity is set to 50 mm/s, reducing the overall execution time w.r.t. the first experiment.

Figure 7.9 shows the measured contour together with the camera and tool paths. The corner is detected correctly. The corner information is used to adapt the execution speed and to set the tool path at the corner. The correct feedforward control at the corner results in the arc shaped paths (for both end effector and actual tool) shown in the figure.

The blow-up at the bottom of Fig. 7.9 clearly indicates the need to correct the measured contour data according to (6.14), which takes the compliance of the camera-setup into account. Although the used test object consists of straight edges, the (uncorrected) measured contour just after the corner is not a straight edge. This is caused by the changing contact situation. As the tool rounds the corner, the direction of the contact force changes over 90° and so does the deformation of the camera-setup. Once this deformation due to the contact force at hand is compensated, the resulting corrected contour is again a straight edge.

Figure 7.9 also shows the measured and the actual tool paths. The measured tool path lies more inwardly than the actual tool path due to the compliance of the tool. To get a good match between the tool path and the desired path (calculated by the vision system), the tool path too needs to be corrected. This correction is similar to the correction of the camera measurement. However, as previously mentioned, the compliance (or stiffness) of the tool (including the wrist) is different from the compliance of the camera-setup.

The maximum allowable approach velocity is 50 mm/s. For higher velocities, it is possible that the camera misses the contour. After all, only a few (center) lines of the image are scanned and processed to limit the needed processing time. If higher velocities than 50 mm/s are desired, the vision system must be instructed to look ahead by scanning the top lines of the grabbed image. If the system does not detect an edge or contour in the top part of the image, a corner is nearby and the tangent velocity has to be reduced to 50 mm/s. Then the image processing won't miss this corner when it passes the center of the image.

7.5 Conclusion

This chapter presents a combined vision/force control approach at corners, being a special case of the planar contour following task.

The vision system is used to watch out for corners. Incorporating the camera/tool deformation in the edge measurements enables an accurate contour

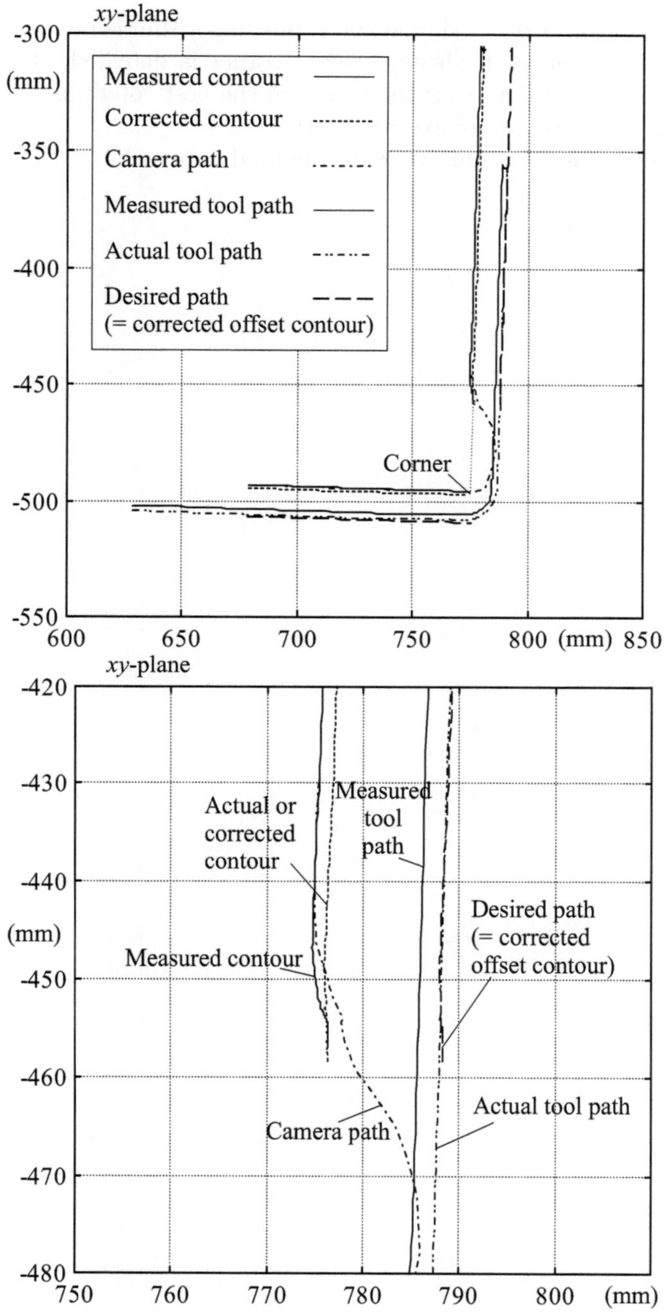

Fig. 7.9. (**top**) Camera path, measured contour and tool path for the second experiment; (**bottom**) Blow-up of the same measurements after the corner

and corner localization. A simple and robust algorithm is implemented to detect corners in the path ahead. Once a corner is detected, the finite state controller is activated to take the corner in the best conditions resulting in a faster and more accurately executed task.

The experimental results validate the used approach.

8

Additional Experiments

8.1 Introduction

This chapter presents experimental results for three combined vision/force tasks. In addition to the results given in the previous chapters, they once more validate our approach.

The first experiment is an example of traded control, according to subtasks 1.3, 1.7 and 1.8 (described in Sect. 4.5). Subtask 1.3 consists of visual aligning the camera with the contour in position and orientation. Subtask 1.7 is a force controlled approach task with fixed task/camera relation. In subtask 1.8, the force controlled contact is maintained (with the tool standing still), while the camera positions over the contour using a variable EOL configuration.

The second experiment is an example of shared control of type 1a (see Fig. 4.3) being a guarded approach.

The third and last experiment uses a fixed EOL configuration. It fully corresponds to the blade polishing task (Task 5b) presented in Fig. 4.8-bottom. However, the test object is a car door.

The next sections describe the experimental setup and present the results for the three experiments[1]. The (corresponding) high level task descriptions were already given in Chap. 4, Sect. 4.5.

8.2 Traded Control in EOL

The first experiment uses the experimental setup of the contour following task of Chap. 6. This is an EOL configuration.

Figure 8.1 shows some of the results for subtasks 1.3, 1.7 and 1.8, described in Sect. 4.5. Vision and force control are stable and the subtasks are all

[1] Preliminary results for Task 5a, the path on surface following task, simulating a truck with semi-trailer, are given by Schillebeeckx and Vandenberk in [72].

executed successfully. Note that subtask 1.7 is based on a fixed EOL configuration. Subtask 1.8, on the other hand, exploits the additional rotational DOF of a variable EOL configuration.

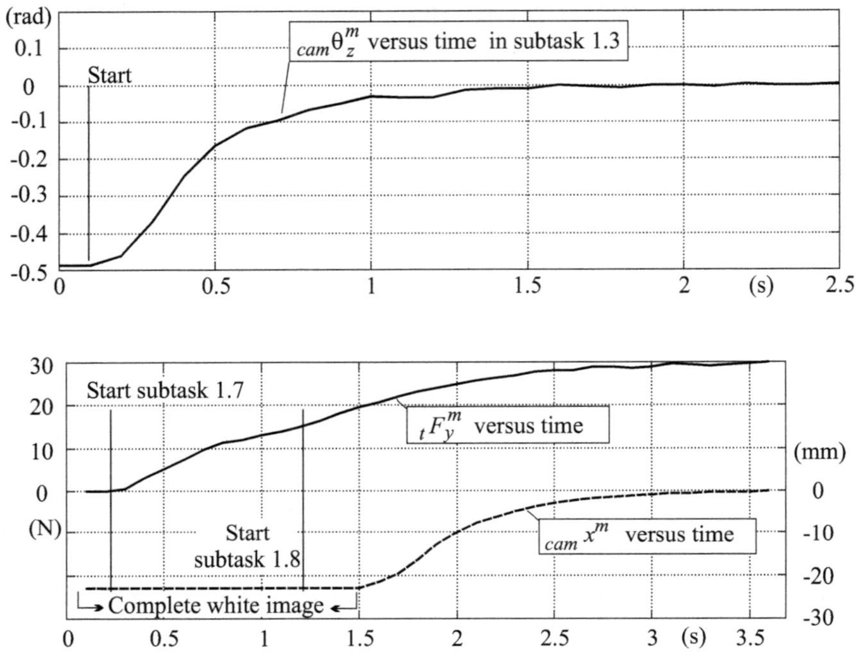

Fig. 8.1. Decay of angular error ${}_{cam}\theta_z^m$ in subtask 1.3 (**top**) and contact force ${}_tF_y^m$ and decay of vision error ${}_{cam}x^m$ in subtasks 1.7 and 1.8 (**bottom**)

8.3 Shared Control in Non-parallel ECL

Figure 8.2 gives the setup for the second experiment. In contrast to the other two, this one is a non-parallel ECL configuration, which is used for a shared control task of type (d^{f+vs}) in an axial direction, according to the first example in Fig. 4.3.

The task goal is to make contact with an object, avoiding excessive contact forces. Since the (exact) position of the object is unknown, a very slow approach velocity is mandatory. In a guarded approach, this approach velocity depends on the difference in measured and desired forces: ${}_tv^f = K^f k^{inv}({}_tF^m - {}_tF^d)$. (With $K^f = 4/\mathrm{s}$, $k^{inv} = 0.075\,\mathrm{mm/N}$ and ${}_tF^d = -30\,\mathrm{N}$, this gives ${}_tv^f = 9\,\mathrm{mm/s}$).

With the vision system, we can speed up the approach simply by augmenting the guarded approach velocity with vision control, proportional to

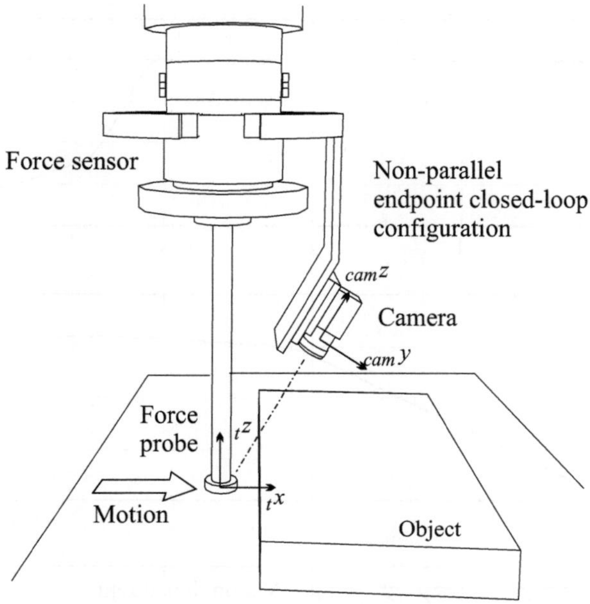

Fig. 8.2. ECL configuration with non-parallel mounting used for shared control in a guarded approach task

the visually measured distance to the object: ${}_t v^{vs} = K^{vs}\ {}^{obj}_{t}x^m$. The total commanded velocity is the sum of the vision and force dependent parts: ${}_t v^c = {}_t v^f + {}_t v^{vs}$.

Figure 8.3 shows the results of the shared control approach experiment. At the start of the task, the object is not visible. The vision dependent velocity is set to a maximum (about 60 mm/s in this experiment but up to 100 mm/s is feasible). Once the object comes in the camera field of view, the vision dependent part of the commanded velocity rapidly decays and the approach slows down. When the distance between the force probe and the object, as seen by the camera, becomes zero, the vision control stops. Due to the chosen setup, there is no contact yet at this point in time. The force dependent control however remains and stably builds up the contact force to -30 N. Evidently, the approach is performed much faster when vision control is added than without vision control, while preserving stability.

In contrast to all other experiments, this experiment utilizes a new camera (SONY XC55bb - NTSC) with square pixels (7.4 by 7.4 µm). The frame rate is 60 Hz. This corresponds to an image frequency equal to 30 Hz. Hence the image sampling period becomes $T^{vs} = 33.3$ ms (instead of 40 ms for the previous PAL camera). Since the robot control and data logging frequency is 100 Hz, the (same) vision control signals are used three or four times in a row. This can be seen from the staircase-like decay of the vision based part in the controlled velocity.

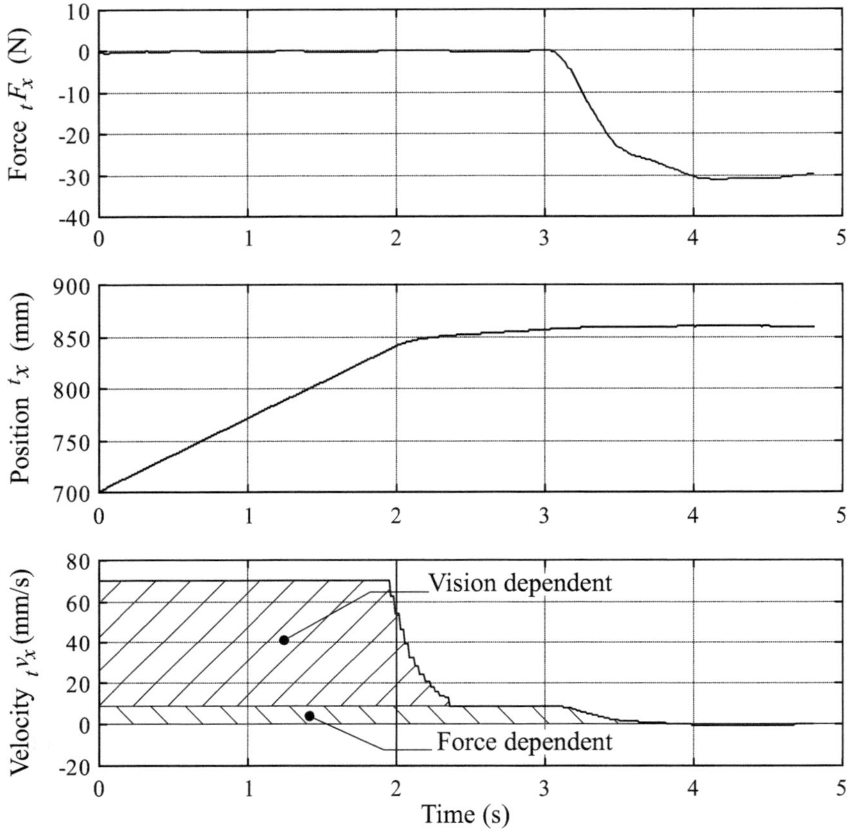

Fig. 8.3. Normal force, absolute task frame position and commanded task frame velocity versus time for the shared control approach experiment

8.4 3D Path in Fixed EOL

The third and last experiment uses a setup with lateral mounted camera as shown in Fig. 8.4. The technical drawing for this mounting is given in appendix Sect. E.4.

This experiment simulates a 'car door polishing' task, which fully corresponds to blade polishing task, being Task 5b in Sect. 4.5.

The task directives are 1) to make contact with the car door and position the tool normal to the car door surface 2) to approach the door edge using vision 3) to align the camera with the edge and 4) to follow the edge. During steps 2 to 4, the force controlled contact of step 1 has to be maintained. Figures 8.5 to 8.7 show the results for steps 2 to 4. Figure 8.5 gives a 3D view and a top view of the paths followed by task and camera frames. As can be seen from the figure, the $_tz$-direction is kept normal to the door surface. This is programmed by demanding zero torque around $_tx$- and $_ty$-directions. Figure 8.6 gives the vision measurements, which are used to control the position in the ${}_t y$-direction and the orientation about the ${}_t z$-axis. At the start of the approach phase, the edge is not yet visible. Once the edge becomes visible, the tool turns to align the camera with the edge. Then, the edge is followed with a constant velocity of 10 mm/s. Figure 8.7 shows the measured contact forces during the experiment. These behave oscillatory due to the very stiff contact between tool and car door. Nevertheless the experiment is executed successfully!

This experiment illustrates that if the image feature of interest is clearly measurable by the vision system and if the tool/object contact incorporates enough compliance (such that an adequate force control is possible), then an integrated vision/force control is feasible.

8.5 Conclusion

This chapter presents additional experimental results for three experiments. All experiments are successfully executed. They show the fitness of the pre-

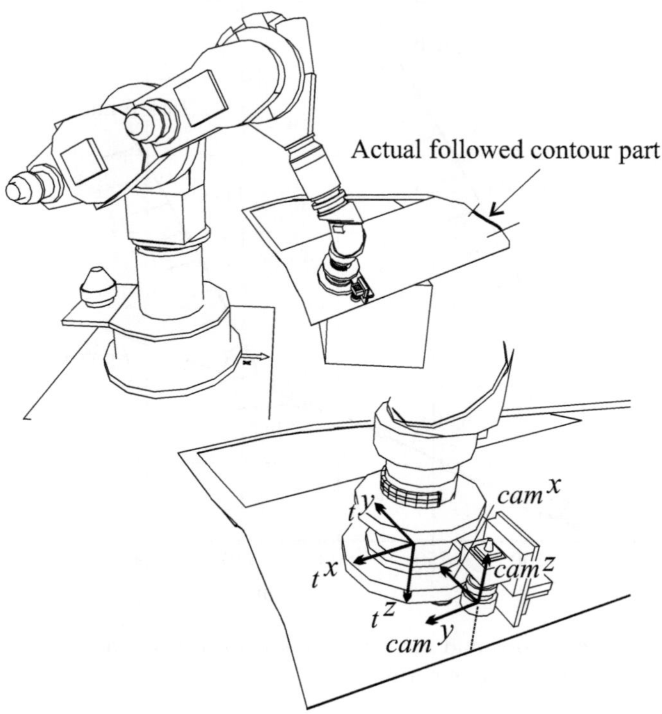

Fig. 8.4. Fixed EOL configuration with a lateral mounted camera for the 'car door polishing' experiment

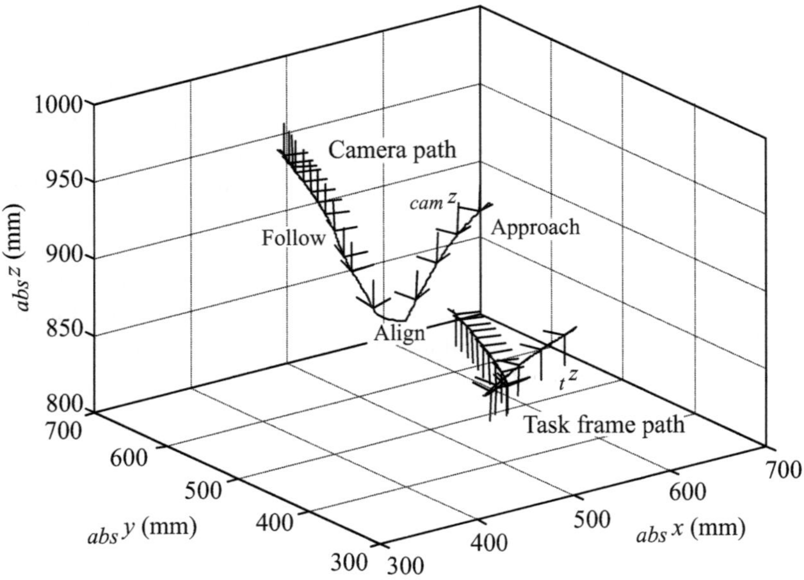

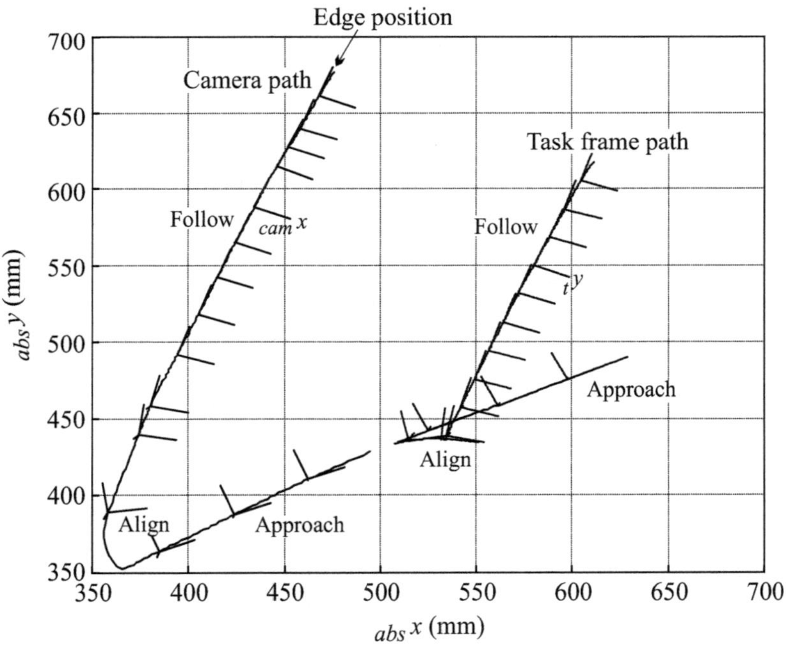

Fig. 8.5. 3D view (**top**) and top view (**bottom**) of task and camera frame paths for the 'car door polishing' experiment

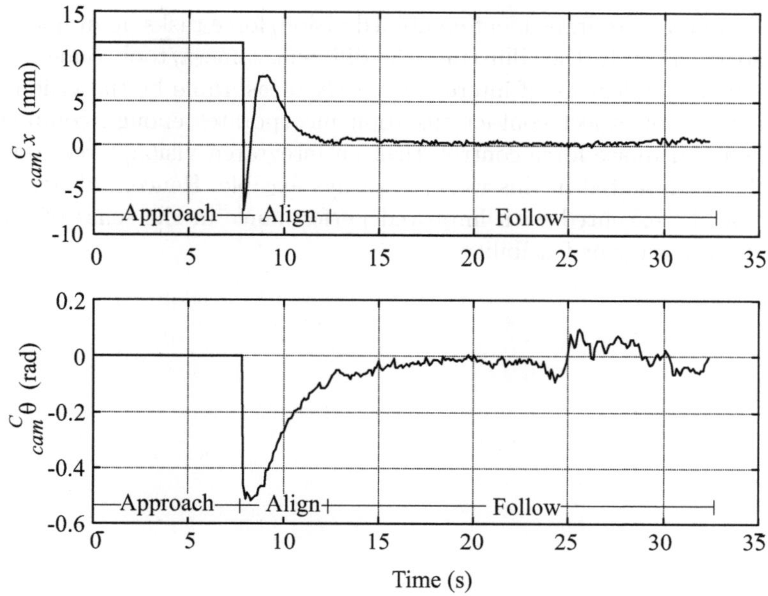

Fig. 8.6. Camera frame errors for the 'car door polishing' experiment

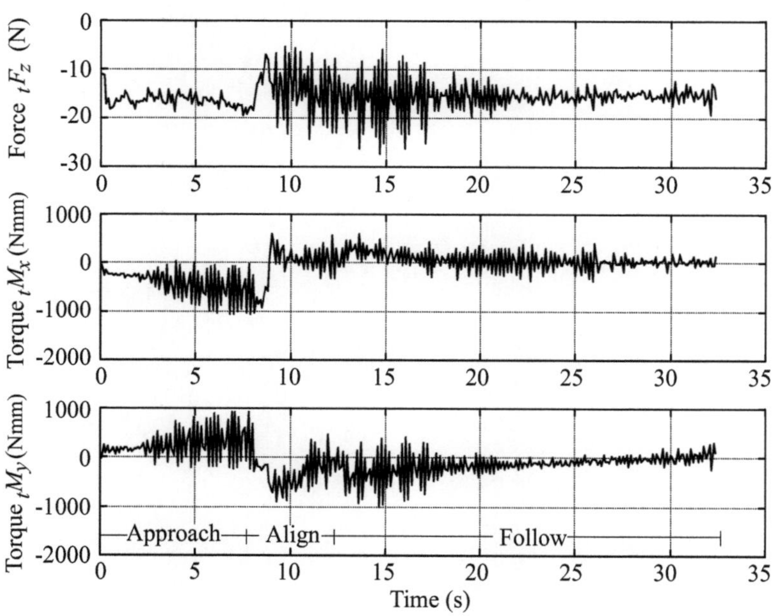

Fig. 8.7. Contact forces for the 'car door polishing' experiment

sented approach to implement combined vision/force tasks in an uncalibrated workspace. They further illustrate the different camera/tool configurations.

If the image feature of interest is clearly measurable by the vision system and if the tool/object contact situation incorporates enough compliance to enable an adequate force control, then an integrated vision/force control approach, as presented in this work, is always feasible. Hence, a wider range of tasks can be executed with improved performance in the sense of increased velocity, accuracy or feasibility.

9

Conclusion

This work presents a *framework to integrate visual servoing and force control* in robotic tasks. Numerous tasks illustrate the fitness of the task frame formalism and the presented hybrid control scheme as a basis for vision and force integration. The *task frame formalism* provides the means to easily model, implement and execute robotic servoing tasks based on combined force control and visual servoing in an *uncalibrated workspace*. The *high level task description* divides the hybrid control space in (separate or mixed) vision and force (among tracking and velocity) subspaces. The key characteristic of the *hybrid controller* is the fusion of vision and force control (possibly with feed-forward) on the basis of velocity set-points, which form the input to the joint controlled robot.

In contrast to what one might expect, the needed effort to accomplish a combined vision/force task (as presented in our approach) is less than the sum of efforts needed for both visual servoing and force control on its own[1]. After all, avoiding collision or depth estimation in a pure visual servoing task or detecting a change in the contact situation, e.g. at a corner, in a force controlled task are annoying operations which are hard to deal with. Using both sensors together remedies these shortcomings. As shown, the vision system is used to look for a starting point, to align with an edge, to measure the path ahead and to watch out for corners in order to speed up the task or to improve the force controlled contact. The force sensor, on the other hand, is used to establish the depth to the object plane, to make and maintain contact with the object and to avoid collisions. Thanks to the calibrated combined mounting of vision and force sensor, the mono vision based computations evidently benefit from the known depth to the object when in contact.

The golden rule of thumb in the above examples is easy: '*Assign specific control directions to that sensor which is best suited to control them*'. This rule implies that a (mono) vision system can only control a maximum of 3 task

[1] Except for the the variable EOL configuration

directions properly. The number of force controlled directions on the other hand depends on the tool/object contact situation (which if not known in advance needs to be identified on-line).

This thesis gives numerous examples of traded, hybrid and shared vision/force controlled tasks. Together with the presented high level task descriptions and the experimental results, they underline the potential of the used approach.

As can be seen from the comparative table 4.1 (in Chap. 4), the key characteristics of any of the presented tasks are included in those which where chosen to be (fully) experimentally tested. We can therefore assume that any task presented in Chap. 4 can effectively be executed.

Also from an *economical* point of view, a combined setup is beneficial. After all, the relatively cheap vision system can improve the performance of the force controlled task, hereby lowering the still existing threshold to use (relatively expensive) force sensors in industrial practice.

9.1 Main Contributions

This thesis contributes to both visual servoing and combined visual servoing/force control approaches. The emphasize lies however on the latter. The main contributions are first grouped into four categories and then discussed in more detail. They are:

A. A new approach to *visual servoing* by using the task frame formalism and relative area algorithms.
B. The **unique** framework for combined vision/force control with (1) *both sensors mounted on the end effector* in (2) a *hybrid vision/force control approach based on the task frame formalism.*
C. Classifications of (1) *shared control* types on the one hand and of (2) *camera/tool configurations* on the other hand. Both classifications apply to combined mounted vision/force control.
D. *Improved planar contour following* (1) for both continuous contours and contours with corners in (2) a *variable endpoint open-loop configuration.*

A. Visual Servoing: Chapter 5 presents a full 3D, image based visual servoing task. The presented (eye-in-hand) visual alignment task is accurately executed incorporating following innovations:

- The task frame formalism is extended to incorporate visual servoing.
- Using the **task frame formalism**, the task goal is translated into control actions in the task frame. Each task frame direction is hereby linked to one specific image feature. By gradually increasing the task control space, the

task is successfully executed, even if the alignment object is only partly visible at the start of the task.

- The image features are all calculated using (partly new) **relative area algorithms** resulting in *real-time*, *robust* control.
- A *rectangular alignment object* is proven to be superior to a circle or an ellipse as alignment object, in the sense that it gives better measurable image features to determine the goal position.

B.1 Combined Vision/Force Mounting: This thesis is one of the first, if not the first, to use an integrated vision/force approach with *both vision and force sensors mounted* on the (same) end effector.

- The poses of vision and force sensors are identified by calibration. Hence, the vision measurements benefit from the known depth to the object when in contact.
- Mounting the camera on the end effector results in a controllable camera position, close to the image feature and hence in a high resolution. Furthermore, aligning the image feature of interest with the optical axis, makes the feature measurement insensitive to the camera/lense distortion.
- With both sensors mounted on the same end effector, the position measurements benefit from the high relative accuracy of the robotic system.
- Incorporating the camera/tool deformation (when in contact) in the measurements enables an accurate contour measurement and corner localization. This improves the match between actual tool path and visually measured contour data.

B.2 Integrated Hybrid Control in the Task Frame Formalism:

- Vision and force sensing are fused in a *hybrid control scheme* which forms the heart of our approach. It consists of a (sensor based) outer control loop around a (joint) velocity controlled robot.
- The *task frame formalism* is the means to easily model, implement and execute combined vision force robotic servoing tasks in an uncalibrated workspace.
- The high level task description divides the hybrid control space in (separate or mixed) **vision**, **force**, **tracking** and **velocity** subspaces. These control subspaces are by definition *orthogonal*. Any control action may be augmented with **feedforward** control.
- The hybrid controller implements *traded*, *hybrid* as well as *shared* (vision/force) control, dependent on the high level task description.

C.1 Shared Control Types: Section 4.4 gives examples of all possible forms of shared control (for both polar and axial directions) in combining vision and force control in one direction or by adding vision based feedforward to either a force direction or a force based tracking direction.

C.2 New Classification of Camera/Tool Configurations: A framework is established to classify combined vision/force tasks, with both sensors mounted on the end effector, into four camera/tool configurations:

1. PARALLEL ENDPOINT CLOSED-LOOP
2. NON-PARALLEL ENDPOINT CLOSED-LOOP
3. FIXED ENDPOINT OPEN-LOOP
4. VARIABLE ENDPOINT OPEN-LOOP

- Examples are given for all configurations but mostly the *EOL configuration with variable task/camera frame relation* is fully explored, since it is the most adequate one for 'simple' image processing and the most challenging in the sense of control.
- All configurations are compatible with the presented hybrid control scheme.

D.1 Improved Planar Contour Following: Chapters 6 and 7 fully examine the planar contour following task for both continuous curves and contours with corners.

- They show how the quality of force controlled planar contour following improves significantly by adding *vision based feedforward* on the *force tracking* direction.
- They are at once an excellent example of shared control.
- An adequate approach is presented to compute the curvature based feedforward on-line and in real-time. The contour is modelled by tangent lines and the curvature is computed as the change in orientation versus arc length. A twice applied least squares solution (once for the tangent, once for the curvature) avoids noisy control signals.
- A simple and robust corner detection algorithm is proposed to adequately detect and measure corners in the path ahead. The corner is rounded, maintaining the contact force, using (again) feedforward. To this end, the control structure is augmented with a finite state machine.

D.2 Variable EOL Control: To realize the improved planar contour following control an variable EOL configuration is suggested.

- An EOL configuration avoids occlusion of the object or contour by the tool and offers the opportunity to detect in advance the path ahead, which, regarded the low vision bandwidth, can only be advantageous.
- For rotationally symmetric tools, a redundancy exists for the rotation in the plane. This degree of freedom is used to position the camera over the contour while maintaining the force controlled contact. This results in a *variable* relation between task and end effector orientations.
- Special control issues arise, due to the time shift between the moment the contour is measured and the moment these data are used. A strategy to match the (vision) measurement data with the contact at hand based on Cartesian position is proposed.

- Using the variable camera/task frame relation, our approach assures a tangent motion for both the task frame (i.e. at the contact point) and the camera frame (i.e. at the intersection of the optical axis with the object plane) along the contour.

9.2 Limitations and Future Work

In view of the limited equipment due to the older type of robot with emerging backlash, limited bandwidth and restricted processing power, the still excellent, achieved results emphasize once more the potential of an outer sensor based control loop around an internal velocity scheme. Better results may be achieved if the sample frequency increases, if the image processing improves (increased computational power and more complex algorithms) or if the controller takes the robot dynamics into account. For the latter, however, the present sample frequency is too low.

Introducing the proposed approach in *industrial practice* faces yet another problem. Although it should be commonly known, as once again underlined in this work, that an outer sensor based control loop is best built around a velocity instead of a position controlled robot, most robot controllers are in fact position based. Furthermore, robot manufactures offer only limited sensor access to their controllers. This complicates the migration of the proposed method to industrial applications.

The presented approach mainly focuses on the lowest level of control. How a given task is best tackled, or with other words, how the control directions are divided over vision and force control by the task description, is left over to the responsibility of the programmer or is subject to *other research* such as programming by human demonstration, active sensing, human-machine interactive programming or task level programming (in which the high level task description is automatically generated and autonomously executed). These latter methods can utilize the presented control approach as a basis to integrate vision and force in their respective fields.

This work uses mono vision only. Future work may look into the specifics of combined *stereo vision* and force control.

Finally, this work is not about vision or *vision processing* by itself. However essential, the vision processing is kept simple and robust, in order to make real-time control possible. With increasing processing power, more complex, real time algorithms will certainly enable the processing of a more diverse range of working scenes. This opens new research possibilities, e.g. by using real-time high-performance snake-algorithms in tracking the image features. The underlying control structure and ideas, however, will remain intact.

A

Derivations

A.1 Frames and Homogeneous Transformations

The position and orientation of (a right-handed) frame a expressed in frame b is:

$$ {}^a_bH = \begin{bmatrix} {}^a_bR & {}^a_bT \\ 000 & 1 \end{bmatrix} \tag{A.1} $$

with ${}^a_bR$ a 3×3 rotation matrix, indicating the orientation of frame a w.r.t. frame b and ${}^a_bT$ a 3×1 translation vector indicating the position of the origin of frame a in frame b.

The position of a point in frame b, being the 3×1 vector ${}_bP$ containing x, y and z coordinates, given the position of this point in frame a being ${}_aP$ follows from

$$ {}_bP = {}^a_bR \, {}_aP + {}^a_bT \;\Leftrightarrow\; {}_b\underline{P} = {}^a_bH \, {}_a\underline{P} \tag{A.2} $$

with $\underline{P}$ the homogeneous coordinates of the point P or

$$ P = \begin{pmatrix} x \\ y \\ z \end{pmatrix} \Leftrightarrow \underline{P} = \begin{pmatrix} x \\ y \\ z \\ 1 \end{pmatrix} . \tag{A.3} $$

${}^a_bH$ thus represents both the homogeneous transformation from frame a to frame b and the pose of frame a in frame b.

Any arbitrary rotation R can be represented by three angles $[\alpha_x, \alpha_y, \alpha_z]'$ indicating the successive rotations about z, y and x-directions:

$$ R = R_z(\alpha_z)\, R_y(\alpha_y)\, R_x(\alpha_x) \tag{A.4} $$

with $R_i(\alpha)$ the rotation matrix for a rotation of α radians about direction i or in particular:

$$R_z(\alpha_z) = \begin{pmatrix} \cos(\alpha_z) & -\sin(\alpha_z) & 0 \\ \sin(\alpha_z) & \cos(\alpha_z) & 0 \\ 0 & 0 & 1 \end{pmatrix},$$

$$R_y(\alpha_y) = \begin{pmatrix} \cos(\alpha_y) & 0 & +\sin(\alpha_y) \\ 0 & 1 & 0 \\ -\sin(\alpha_y) & 0 & \cos(\alpha_y) \end{pmatrix}, \tag{A.5}$$

$$R_x(\alpha_x) = \begin{pmatrix} 1 & 0 & 0 \\ 0 & \cos(\alpha_x) & -\sin(\alpha_x) \\ 0 & \sin(\alpha_x) & \cos(\alpha_x) \end{pmatrix}.$$

The computation of the rotation matrix from the rotation angles using (A.4) and (A.5) then comes downs to (with c = cos and s = sin):

$$R = \begin{pmatrix} c(\alpha_z)c(\alpha_y) & c(\alpha_z)s(\alpha_y)s(\alpha_x) - s(\alpha_z)c(\alpha_x) & c(\alpha_z)s(\alpha_y)c(\alpha_x) + s(\alpha_z)s(\alpha_x) \\ s(\alpha_z)c(\alpha_y) & s(\alpha_z)s(\alpha_y)s(\alpha_x) + c(\alpha_z)c(\alpha_x) & s(\alpha_z)s(\alpha_y)c(\alpha_x) - c(\alpha_z)s(\alpha_x) \\ -s(\alpha_y) & c(\alpha_y)s(\alpha_x) & c(\alpha_y)c(\alpha_x) \end{pmatrix} \tag{A.6}$$

A.2 Screw Transformations

Let ${}_aV$ be a 6×1 generalized velocity vector (also called twist or velocity screw), containing 3 axial and 3 polar velocities, expressed in frame a:

$${}_aV = [\ {}_av_x,\ {}_av_y,\ {}_av_z,\ {}_a\omega_x,\ {}_a\omega_y,\ {}_a\omega_z]'. \tag{A.7}$$

Let ${}^a_bS^v$ be the velocity screw transformation for the recalculation of a velocity screw in frame a to its equivalent[1] velocity screw in frame b:

$${}_bV = {}^a_bS^v\ {}_aV. \tag{A.8}$$

Using the relative pose of frame a w.r.t. frame b, ${}^a_bH$, as introduced in Sect. A.1, the velocity screw transformation then corresponds to

$${}^a_bS^v = \begin{bmatrix} {}^a_bR & {}^a_bR\,{}^a_bT\times \\ Zeros & {}^a_bR \end{bmatrix} \tag{A.9}$$

with $Zeros$ a 3×3 matrix with all elements equal to zero and with

$${}^a_bT\times = \begin{pmatrix} 0 & -{}^a_bT_z & {}^a_bT_y \\ {}^a_bT_z & 0 & {}^a_bT_x \\ -{}^a_bT_y & {}^a_bT_x & 0 \end{pmatrix}. \tag{A.10}$$

[1] for an identical rigid body motion

Equations A.8 and A.9 can also be subdivided into

$$\begin{pmatrix} {}_bv_x \\ {}_bv_y \\ {}_bv_z \end{pmatrix} = {}^a_bR \left(\begin{pmatrix} {}_av_x \\ {}_av_y \\ {}_av_z \end{pmatrix} + {}^a_bT \times \begin{pmatrix} {}_a\omega_x \\ {}_a\omega_y \\ {}_a\omega_z \end{pmatrix} \right) \tag{A.11}$$

and

$$\begin{pmatrix} {}_b\omega_x \\ {}_b\omega_y \\ {}_b\omega_z \end{pmatrix} = {}^a_bR \begin{pmatrix} {}_a\omega_x \\ {}_a\omega_y \\ {}_a\omega_z \end{pmatrix} . \tag{A.12}$$

Let ${}_aW$ be a 6×1 generalized force vector (also called wrench or force screw), containing 3 forces and 3 moments, expressed in frame a:

$${}_aW = [\ {}_aF_x,\ {}_aF_y,\ {}_aF_z,\ {}_aM_x,\ {}_aM_y,\ {}_aM_z]'. \tag{A.13}$$

Let ${}^a_bS^f$ be the force screw transformation for the recalculation of a wrench in frame a to its equivalent[2] wrench in frame b:

$${}_bW = {}^a_bS^f\ {}_aW. \tag{A.14}$$

The force screw transformation then corresponds to

$${}^a_bS^f = \begin{bmatrix} {}^a_bR & Zeros \\ {}^a_bR\,{}^a_bT\times & {}^a_bR \end{bmatrix} \tag{A.15}$$

with $Zeros$ and ${}^a_bT\times$ as defined above.

A.3 A Denavit-Hartengberg Representation of the KUKA361 Robot

The Denavit-Hartenberg (DH) representation [37] is an easy formalism to describe the kinematic structure of a serial manipulator. Each link of the manipulator is associated with a frame, going from the absolute world frame (frame 0) to the end effector frame (frame 6 for a 6 degree of freedom (DOF) manipulator). The transition of frame $(i-1)$ to frame (i) corresponds to either a translation DOF or a rotation DOF in the connection of link $(i-1)$ and link (i). The KUKA361 robot has 6 rotation DOF. Using the DH representation the pose of the end effector (frame) is calculated as a function of the robot joint values $q = [q_1 \dots q_6]'$. This is called the forward kinematic transformation $FWK(q)$.

[2] for an identical force equilibrium

A.3.1 Formalism

The DH formalism describes the transition from frame $(i-1)$ to frame (i) with four parameters:

α_i the angle between the z_{i-1} and the z_i-axes about the x_i-axis;
a_i the distance between the z_{i-1} and the z_i-axes along the x_i-axis;
d_i the distance from the origin of frame $i-1$ to the x_i-axis along the z_{i-1}-axis;
θ_i the angle between the x_{i-1} and the x_i-axes about the z_{i-1}-axis, which normally includes the variable joint value q_i.

Frame (i) thus follows from frame $(i-1)$ by the following four steps:

1. Rotate frame $(i-1)$ over the angle θ_i around the z_{i-1}-axis.
2. Translate the frame along the z_{i-1}-axis by the distance d_i.
3. Translate the frame along the x_i-axis by the distance a_i.
4. Rotate the frame over the angle α_i around the x_i-axis.

Table A.1. DH parameters for KUKA361 manipulator with distances in mm and angles in rad

i	1	2	3	4	5a	5b	6
α_i	$-\pi/2$	0	$\pi/2$	$\pi/6$	$-\pi/3$	$\pi/6$	0
a_i	0	480	0	0	0	0	0
d_i	1020	0	0	645	0	0	120
θ_i	q_1	$q_2-\pi/2$	$q_3+\pi/2$	$q_4+\pi/2$	q_5	$-q_5$	$q_6-\pi/2$

These four steps correspond to the homogeneous transformation ${}_{i-1}^{\ \ i}H$, being

$$
{}_{i-1}^{\ \ i}H = \begin{bmatrix} \cos\theta_i & -\sin\theta_i\cos\alpha_i & \sin\theta_i\sin\alpha_i & a_i\cos\theta_i \\ \sin\theta_i & \cos\theta_i\cos\alpha_i & -\cos\theta_i\sin\alpha_i & a_i\sin\theta_i \\ 0 & \sin\alpha_i & \cos\alpha_i & d_i \\ 0 & 0 & 0 & 1 \end{bmatrix}. \tag{A.16}
$$

A recursive application of this transformation results in

$$
{}_{abs}^{\ ee}FWK(q) = {}_{abs}^{\ \ 1}H(q_1)\,{}_1^2H(q_2)\ldots{}_5^{ee}H(q_6). \tag{A.17}
$$

A.3.2 KUKA361

Figure A.1 gives an overview of the DH frames for the KUKA361 manipulator. The figure is drawn with all joint values $[q_1 \ldots q_6]'$ equal to zero. Table A.1

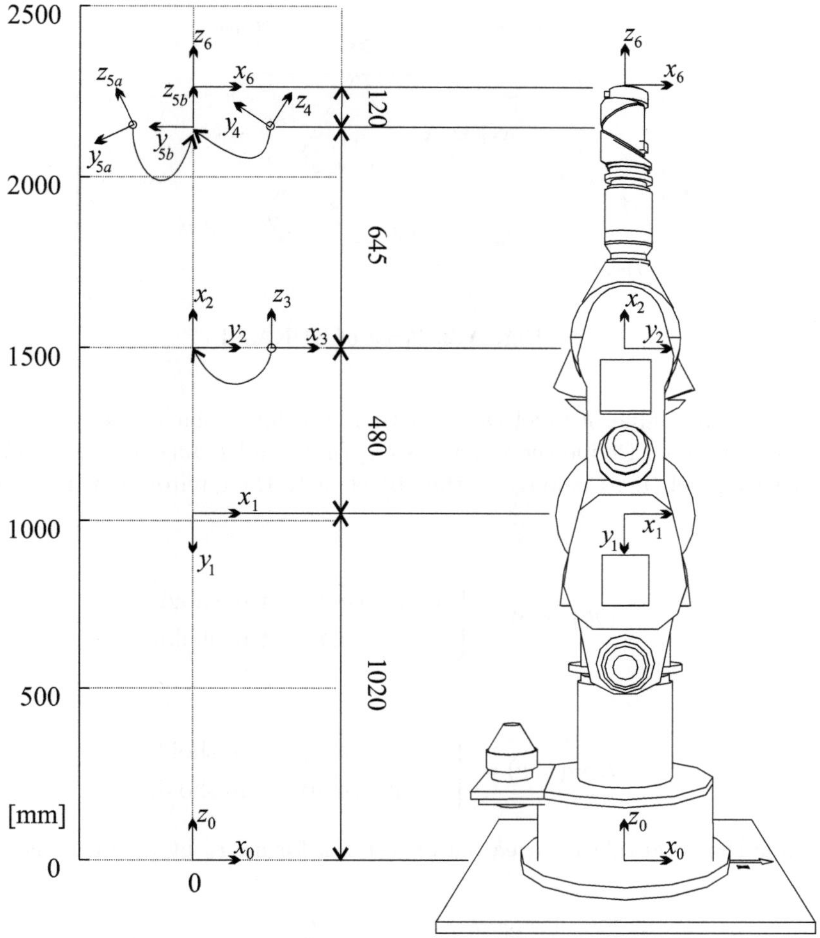

Fig. A.1. DH frames of KUKA361 manipulator

summarizes the DH parameters for the KUKA361[3]. Because joint 5 consists of two moving parts, as shown in Fig. A.2, one additional frame is needed, hereby splitting frame 5 in a and b. The rotations around the axes $5a$ and $5b$ in Fig. A.2, however, are always equal with opposite signs.

A.4 Point Symmetry with Relative Area Method

Assume that the object is dark and the environment is bright. Assume further that the object lies in the left part of the image. Let $I(u, v)$ be the intensity

[3] Practical note: to get an identical robot position for a given set of joint values with this representation as with COMRADE, the joint values (q_4, q_5, q_6) have to be inverted.

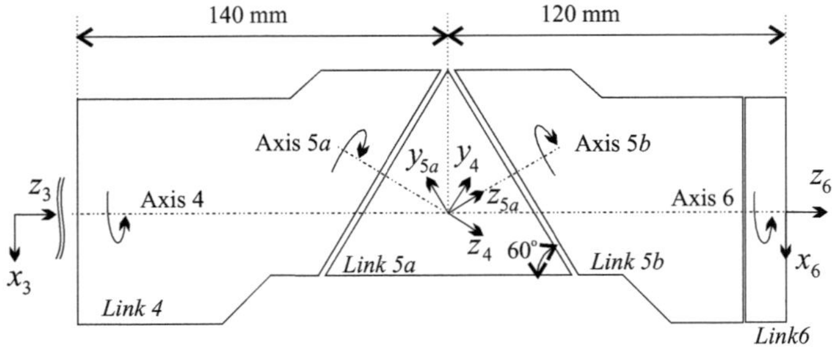

Fig. A.2. Wrist of KUKA361

or grey-level value of pixel (u, v), with $I = 0$ for a black pixel and $I = 255$ for a white one. Define the functions $Obj(u, v)$ and $Env(u, v)$, which indicate whether pixel (u, v) belongs to the object or to the environment respectively, as

$$Obj(u, v) = \begin{cases} 1 & \text{if } I(u, v) < \text{threshold}, \\ 0 & \text{if } I(u, v) \geq \text{threshold}, \end{cases} \tag{A.18}$$

and

$$Env(u, v) = \begin{cases} 0 & \text{if } I(u, v) < \text{threshold}, \\ 1 & \text{if } I(u, v) \geq \text{threshold}. \end{cases} \tag{A.19}$$

Define the total relative area parameter T_{ra} for an x_s by y_s image as

$$T_{ra} = \left[\sum_{u,v=1,1}^{x_s,y_s} Obj(u, v) - \sum_{u,v=1,1}^{x_s,y_s} Env(u, v) \right] \tag{A.20}$$

Then, the distance ${}^{c}x_p$, defined in figure 3.21, is given by (equivalent to (3.15)):

$${}^{c}x_p = \frac{1}{2y_s} T_{ra} \tag{A.21}$$

Define the y-signed relative area parameter YS_{ra} as

$$YS_{ra} = \sum_{u=1}^{x_s} \left[\sum_{v=1}^{y_s/2} Obj(u, v) - \sum_{v=\frac{y_s}{2}+1}^{y_s} Obj(u, v) \right]. \tag{A.22}$$

Then, the angle ${}^{c}\theta$, defined in figure 3.21, is (equivalent to (3.16))

$${}^{c}\theta = -\arctan \left[\frac{4}{y_s^2} YS_{ra} \right]. \tag{A.23}$$

Finally, define the shifted x-signed relative area parameter SXS_{ra} as

$$SXS_{ra} = -\sum_{v=1}^{y_s} \left[\sum_{u=1}^{\text{floor}(\frac{x_s}{2}+{}^C x_p)} Obj(u,v) - \sum_{\text{floor}(\frac{x_s}{2}+{}^C x_p)+1}^{x_s} Obj(u,v) \right]. \quad (A.24)$$

The straightness of a vertical contour in the image can now be checked by verifying the point symmetry of the image w.r.t. a frame centered at the contour. This is a necessary but not sufficient condition.

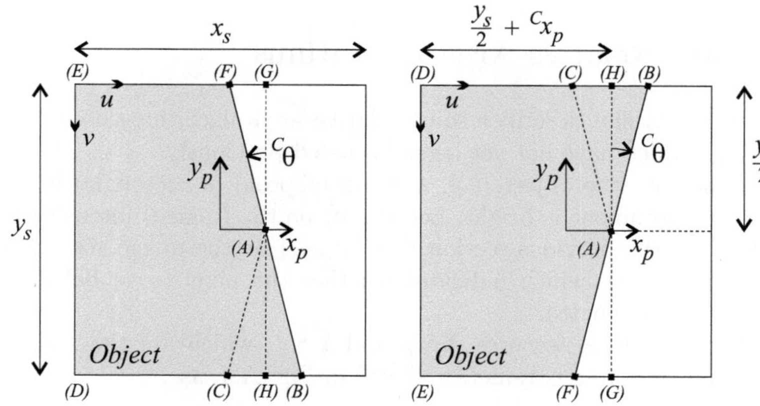

Fig. A.3. Illustration for point symmetry calculation for images of a contour with positive angle (**left**) and negative angle (**right**)

Consider the two situations shown in Fig. A.3. The absolute value of parameter SXS_{ra} is equal to the area of $ACDEF$. The absolute value of YS_{ra} is equal to the area of ABC. The sum of both areas is equal to the area $DEGH$. For a contour with positive angle (Fig. A.3-left), both parameters SXS_{ra} and YS_{ra} are negative, thus

$$SXS_{ra} + YS_{ra} = -\frac{(x_s + 2\,{}^C x_p)\, y_s}{2}. \quad (A.25)$$

For a contour with negative angle (Fig. A.3-right), SXS_{ra} is negative and YS_{ra} is positive, thus

$$SXS_{ra} - YS_{ra} = -\frac{(x_s + 2\,{}^C x_p)\, y_s}{2}. \quad (A.26)$$

Both (A.25) and (A.26) together give following theorem:

$$\text{If an image with vertical contour is point symmetric, then} \quad SXS_{ra} - |YS_{ra}| = -\tfrac{(x_s + 2\,{}^C x_p)\,y_s}{2}. \tag{A.27}$$

This equation is equivalent to (3.17). Due to rounding off errors and the spatial quantization of the image,(A.27) is hardly ever fulfilled. Hence, the image is regarded as point symmetric if

$$\left| SXS_{ra} - |YS_{ra}| + \frac{(x_s + 2\,{}^C x_p)\,y_s}{2} \right| < threshold_1. \tag{A.28}$$

A good value for $threshold_1$ is $2\,y_s$.

A.5 Some Relative Area Algorithms

This section briefly describes some relative area algorithms used to process the image, which have not yet been discussed previously.

Assume that the object (e.g. a circle, ellipsoid or rectangle) is dark and that the environment is bright. Let $I(u,v)$ be the (image) intensity function as defined in the previous section. Let (x_s, y_s) be the image size. Define the function $Obj(u,v)$, which indicates whether the pixel (u,v) belongs to the object according (A.18).

Define the image features XS_{ra} and YS_{ra}, which are the x-signed and y-signed relative area parameters shown in Fig. A.4, as

$$XS_{ra} = -\sum_{v=1}^{y_s} \left[\sum_{u=1}^{x_s/2} Obj(u,v) - \sum_{u=\frac{x_s}{2}+1}^{x_s} Obj(u,v) \right] \tag{A.29}$$

and

$$YS_{ra} = \sum_{u=1}^{x_s} \left[\sum_{v=1}^{y_s/2} Obj(u,v) - \sum_{v=\frac{y_s}{2}+1}^{y_s} Obj(u,v) \right]. \tag{A.30}$$

Then the distances ${}^{obj}x$ and ${}^{obj}y$, which are the coordinates (pix) of the center of the object, as shown in Fig. A.4, are proportional to respectively XS_{ra} and YS_{ra}. If the image feature XS_{ra} is zero, then also ${}^{obj}x$ will be zero and likewise for YS_{ra} and ${}^{obj}y$.

Define the image features XYS_{ra}, which gives the difference in object pixels lying on the one hand in the first and third quadrant and on the other hand in the second and fourth quadrant of the image, as

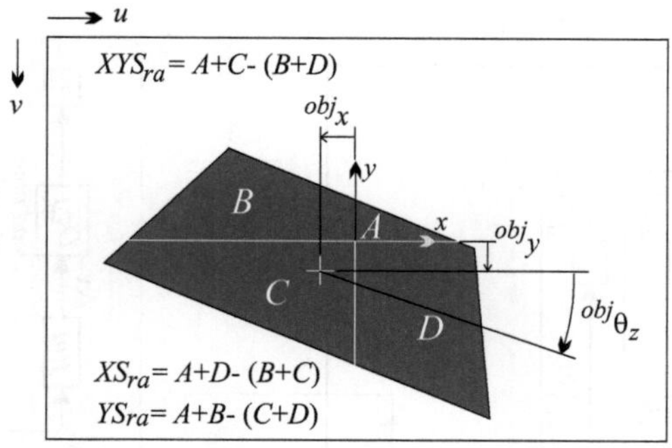

Fig. A.4. Relative area parameters: XS_{ra}, YS_{ra} and XYS_{ra}

$$\begin{aligned} XYS_{ra} = &- \sum_{u=1}^{x_s/2} \sum_{v=1}^{y_s/2} Obj(u,v) - \sum_{u=\frac{x_s}{2}+1}^{x_s} \sum_{v=\frac{y_s}{2}+1}^{y_s} Obj(u,v) \\ &+ \sum_{u=1}^{x_s/2} \sum_{v=\frac{y_s}{2}+1}^{y_s} Obj(u,v) + \sum_{u=\frac{x_s}{2}+1}^{x_s} \sum_{v=1}^{y_s/2} Obj(u,v). \end{aligned} \tag{A.31}$$

Then the angle $^{obj}\theta_z$, as shown in Fig. A.4, is proportional to the image feature parameter XYS_{ra}. If the image feature XYS_{ra} is zero, then also $^{obj}\theta_z$ will be zero.

A.6 2D Arc Following Simulation

Consider a task consisting of a rotating task frame following an arc shaped contour. The objectives for this task are:

- move with constant $_ty$-velocity;
- keep the task frame tangent to the contour;
- minimize the position error between the task frame origin and the contour along the $_tx$-direction.

This section describes the SIMULINK/MATLAB® control scheme, given in Fig. A.5, including the system, the sensor and the controller. This model simulates the dynamic behaviour of the task frame for the arc following task.

A.6.1 The System

The system to be controlled is the 2D pose of the task frame. The input are the commanded relative velocities ($_tv_x^c$, $_tv_y^c$ and $_t\omega_z^c$) of the task frame. The output of the system is the 2D absolute pose ($_{abs}^{\ \ t}x(t)$, $_{abs}^{\ \ t}y(t)$, $_{abs}^{\ \ t}\theta_z(t)$) of the

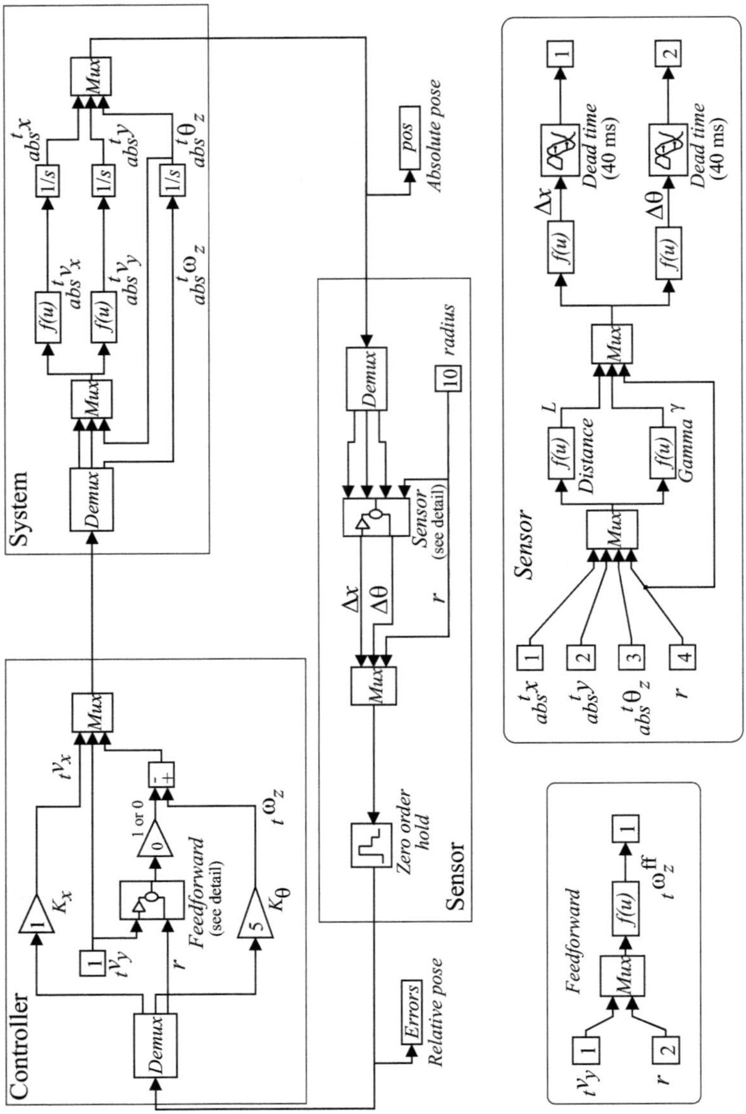

Fig. A.5. Simulation scheme for computing the dynamic behaviour of the tracking errors in following an arc with a rotating task frame task frame at any time instant t. The computation of the absolute pose from the relative velocities is equal to the multiplication of the input velocity vector with a rotation matrix, followed by an integration from velocity to position:

$$s \begin{pmatrix} {}^{t}_{abs}x \\ {}^{t}_{abs}y \\ {}^{t}_{abs}\theta_z \end{pmatrix} = \begin{pmatrix} \cos({}^{t}_{abs}\theta) & -\sin({}^{t}_{abs}\theta) & 0 \\ \sin({}^{t}_{abs}\theta) & \cos({}^{t}_{abs}\theta) & 0 \\ 0 & 0 & 1 \end{pmatrix} \begin{pmatrix} {}_{t}v_x^c \\ {}_{t}v_y^c \\ {}_{t}\omega_z^c \end{pmatrix} . \tag{A.32}$$

Figure A.6 illustrates the setup.

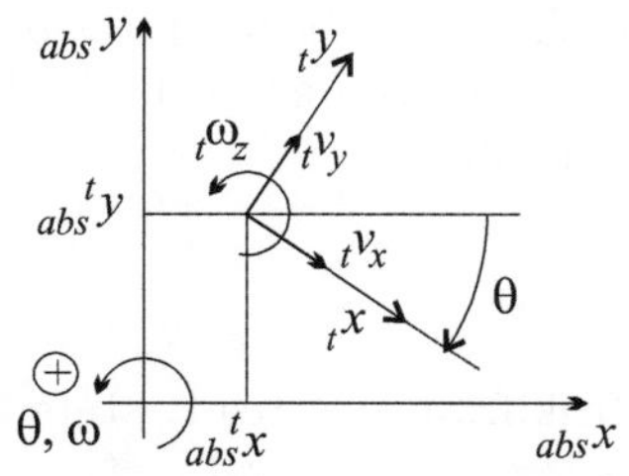

Fig. A.6. Task frame pose w.r.t. the absolute frame

A.6.2 The Sensor

The sensor 'measures' the position and orientation errors (Δx and $\Delta\theta$) of the contour relative to the task frame. For the arc shaped contour of Fig. A.7, these errors are

$$\begin{cases} \Delta x = L\cos\gamma - \sqrt{r^2 - L^2\sin^2\gamma} \\ \Delta\theta = -\arcsin\left(\frac{L sin \gamma}{r}\right) \end{cases} \tag{A.33}$$

with

$$\begin{cases} L = \sqrt{{}^{t}_{abs}y^2 + (r - {}^{t}_{abs}\, x)^2} \\ \gamma = \frac{\pi}{2} + {}^{t}_{abs}\theta - \arctan\left(\frac{r - {}^{t}_{abs}x}{{}^{t}_{abs}y}\right) \end{cases} . \tag{A.34}$$

γ is the angle between the ${}_{t}x$-axis (line AD) and the line (AC), connecting the center of the arc with the origin of the task frame, as shown in Fig. A.7; r is the radius of the arc; L is the distance between the center of the arc (point C) and the origin of the task frame (point$A$).

The sensor includes a time delay of 40 ms and a zero order hold circuit, as shown in the detail of the sensor in Fig. A.5, to simulate the vision sensor which is a discrete system that has to operate at the video frame rate of 25 Hz.

A.6.3 The Controller

The equations for the proportional control of ($_t x$) position and orientation of the task frame, possibly with feedforward, are

$$\begin{cases} {}_t\omega_z^{\text{c}} = K_\theta \Delta\theta + {}_t\omega_z^{\text{ff}} \\ {}_t v_x^{\text{c}} = K_x \Delta x \end{cases} . \tag{A.35}$$

A.6.4 Parameters

The simulation allows for variation in several parameters such as: the control constants (K_x and K_θ), the begin pose of the task frame (${}^{t}_{abs}x(0)$, ${}^{t}_{abs}y(0)$, ${}^{t}_{abs}\theta_z(0)$), the radius of the arc (r), the forward velocity (${}_t v_y^c$) and the use of feedforward control.

A.6.5 Some Results

The actual path of the task frame in steady state follows from three simultaneous, fixed velocities, being ${}_t v_y$, ${}_t v_x^{\text{ss}}$ and ${}_t\omega^{\text{ss}}$:

$${}_t v_x^{\text{ss}} = K_x \Delta x^{\text{ss}} \approx \frac{{}_t v_y^2}{r K_\theta} \quad \text{and} \quad {}_t\omega^{\text{ss}} = K_\theta \Delta\theta^{\text{ss}} = \frac{-{}_t v_y}{r}. \tag{A.36}$$

This results in a task frame motion which consists of an instantaneous rotation (${}_t\omega^{\text{ss}}$) and a translation (${}_t v_x^{\text{ss}}$). The center of the rotation lies at a distance r (which is also the radius of the arc) along the ${}_t x$-axis. At the same time, the task frame translates in the ${}_t x$-direction. This implies that the center of the instantaneous rotation, itself moves on a circular path with radius r_o around the center of the arc. r_o equals $-{}_t v_x^{\text{ss}} / {}_t\omega^{\text{ss}} = {}_t v_y / K_\theta$. See Fig. A.8. Other simulation results are given in Figs. 3.10 and 3.11 in Sect. 3.2.

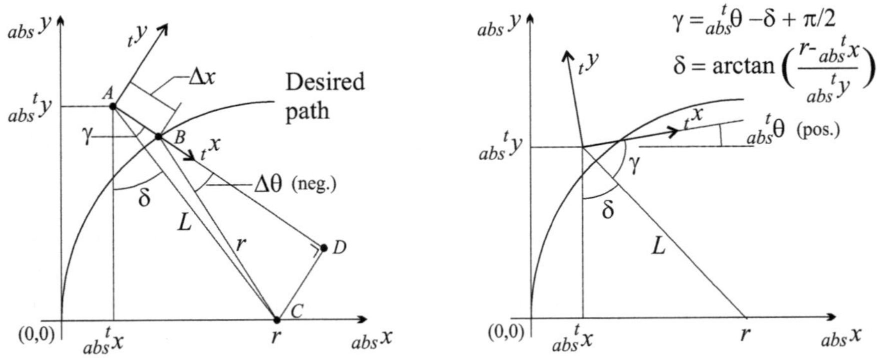

Fig. A.7. Illustration for the computation of position and tracking errors, Δx and $\Delta\theta$, of the arc relative to an arbitrary task frame pose (**left**) and computation of γ (**right**)

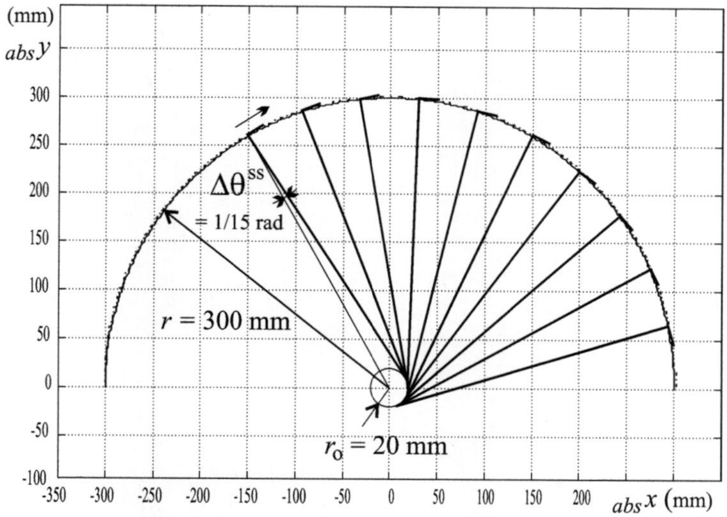

Fig. A.8. The center of the instantaneous rotation of the task frame follows itself a circular path around the center of the arc; the settings are $K_x = 1$ (1/s), $K_\theta = 1$ (1/s), ${}_tv_y = 20$ (mm/s) and no feedforward

B

Contour Fitting

This appendix presents several contour fitting models and evaluates them. Important evaluation criteria are the accuracy and robustness of the fitted function w.r.t. the real contour, the computational burden in obtaining the model parameters and the suitability of the model as a basis for the control data computation. The input for the contour fitting is the set of n contour points, given by $[X_p, Y_p]$ with $X_p = [x_p^1 \dots x_p^n]'$ and $Y_p = [y_p^1 \dots y_p^n]'$. Normally, the contour lies 'top-down' in the image. Hence it is logical to represent the contour with x as a function of y. For reasons of convenience the index p is left away for the moment. Following contour fitting models are described and discussed in this chapter:

- a set of tangents,
- a full second order function (circle, ellipse, parabola or hyperbola),
- an interpolating polynomial,
- a third order non-interpolating polynomial ,
- interpolating cubic splines and
- the parameterized version of the latter two.

All these models, with some variants, are implemented in a Contour Fitting Graphical User Interface (CFGUI), under MATLAB®, enabling a user friendly qualitative comparison of the contour fitting models. Section B.8 briefly describes the functionality of this CFGUI.

B.1 The Tangent Model

The simplest contour model is a line:

$$x = ay + b. \tag{B.1}$$

The model parameters a and b follow from a least squares fit of the model through the set of contour points. If the contour points lie closely together,

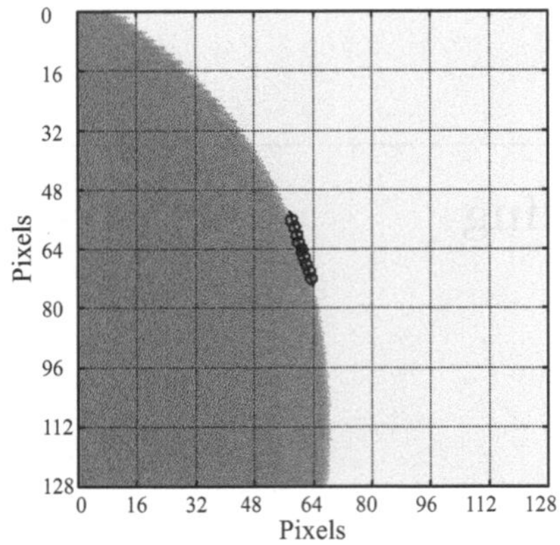

Fig. B.1. Tangent to contour

this fitted line will approximate the tangent to the contour at the center of the data set. Figure B.1 gives an example. The main advantage of this model is it simplicity. The contour pose (position ${}^{c}x_p$ and orientation ${}^{c}\theta$) for one single point on the contour is directly given by the model. When the contour pose is given for consecutive points on the contour, the curvature κ can be computed as the change in orientation ${}^{c}\theta$ as a function of the arc length s:

$$\kappa = \frac{d\,{}^{c}\theta}{ds}. \tag{B.2}$$

In practice, an approximation for κ follows from the least squares solution of a set of m first order equations

$${}^{c}\theta(i) = \kappa \cdot s(i) + cte \quad i = 1 \ldots m. \tag{B.3}$$

The m (e.g. 9) pairs $({}^{c}\theta(i), s(i))$ lie symmetrically around the position of interest. Figure B.2 gives an example. The least squares method has an averaging effect on the computed curvature. This is an important advantage. The curvature is, after all, proportional to the second derivative of the position. The curvature computation is thus a very noise sensitive operation. Using a least squares solution avoids the buildup of noise. The computed value is the average curvature over a finite line segment. This implies also the levelling out of the curvature profile at contour segments with extreme or sudden change in curvature. The optimal arc length, over which the average curvature is computed (see Fig. B.2), thus follows from a trade off between noise reduction and signal deformation.

B.2 Full Second Order Function

Often used as contour models (e.g. [33, 47, 71]) are circles and ellipses. They are described by following equation (in the variables x, y):

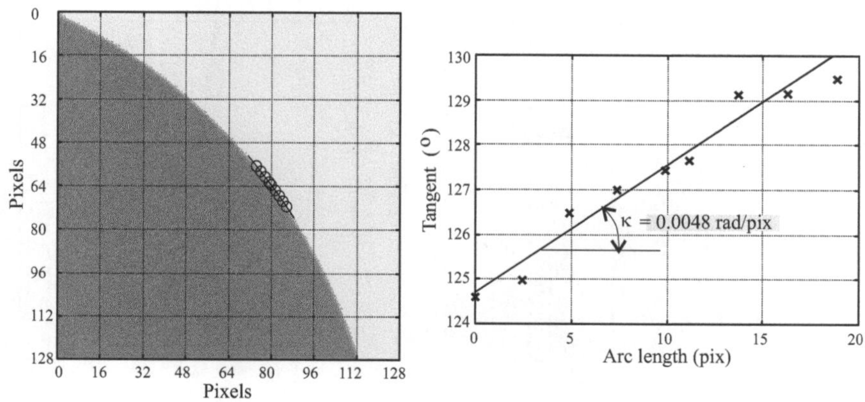

Fig. B.2. Tangent to contour in nine points(**left**); corresponding least squares solution for the curvature computation (**right**)

$$ax^2 + bxy + cy^2 + dx + ey + 1 = 0. \tag{B.4}$$

Depending on the values of the parameter set (a, b, c, d, e), this equation may also represent a parabola or hyperbola. If the data set of contour points is known to lie on a circle or ellipse, this equation will result in a good (least squares) fit of model and likewise control data (see Fig. B.3-left). If, on the other hand, the contour contains a deflection point (or is not circular), the resulting fit may be worthless. Fitting the contour points by a full second order function lacks robustness. Figure B.3-right gives an example: a good (total least squares) hyperbolic fit deters completely by changing only one point by a pixel. Hence, with arbitrary contours a full second order contour model is no option.

B.3 Interpolating Polynomial

If the contour points are correct, the analytical curve must pass trough each point. Interpolating polynomials fulfill this desire. They are described by:

$$x = a_{n-1}y^{n-1} + a_{n-2}y^{n-2} + \ldots + a_1y + a_0 \tag{B.5}$$

with n the number of points. Interpolating polynomials with $n > 3$, however, oscillate. They neither preserve the sign, nor the monotonicity, nor the convexity of the contour data set. They are therefore also rejected.

B.4 Third Order Non-interpolating Polynomial

Non-interpolating polynomials remedy most of the previously mentioned disadvantages of interpolating polynomials. They use more points to be fitted

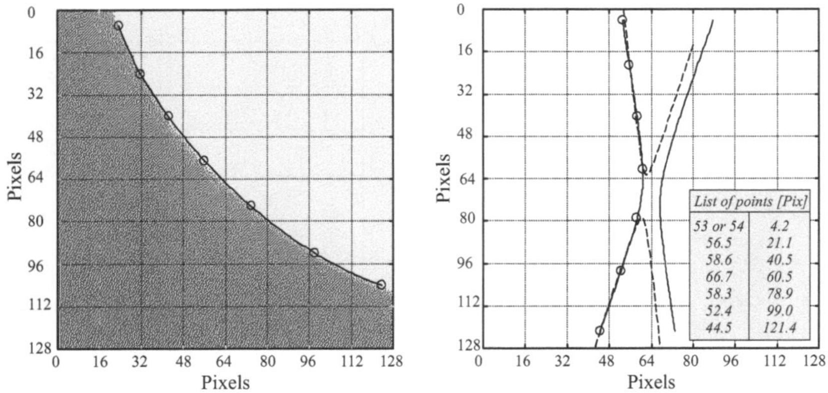

Fig. B.3. Least squares fits of full second order function through selected contour points: (**left**) good elliptic fit; (**right**) changing only one point (away from the center of the image) by a pixel causes a good hyperbolic fit to deter in a useless one

closely to but not exactly through. The order of the polynomial is commonly limited to three:

$$x = ay^3 + by^2 + cy + d. \tag{B.6}$$

The parameters (a, b, c, d) again follow from a least squares fit. Advantageous to this method is the smoothing effect of the least squares fit which reduces oscillations and suppresses noise. Moreover, the order of the contour points is not important. The control data are easily derived: e.g. the curvature of the contour at position $x = 0$ is $\kappa = (-2b)/(1 + c^2)^{\frac{3}{2}}$. The accuracy of the polynomial fit, however, is not satisfactory as illustrated in Fig. B.4. With increasing curvature, the accuracy of the polynomial fit decreases. Figure B.7 gives an example of the computed curvature profile with a polynomial fit through a set of points lying on a arc shaped contour.

B.5 Interpolating Cubic Splines

Interpolating[1] splines (or p-splines) are a set of $n - 1$ functions s_k which are locally defined over the range $[y_k, y_{k+1}]$, for $k = 1, \ldots, n$-1. They are continuous in the contour points, which are called the knots, and satisfy the interpolation condition

$$\begin{cases} s_k(y_k) = x_k, \\ s_k(y_{k+1}) = x_{k+1}, \end{cases} \quad k = 1, \ldots, n{-}1. \tag{B.7}$$

[1] Non-interpolating splines (e.g. B-splines) are not considered since the knots are assumed to be correct contour points and thus need to be interpolated.

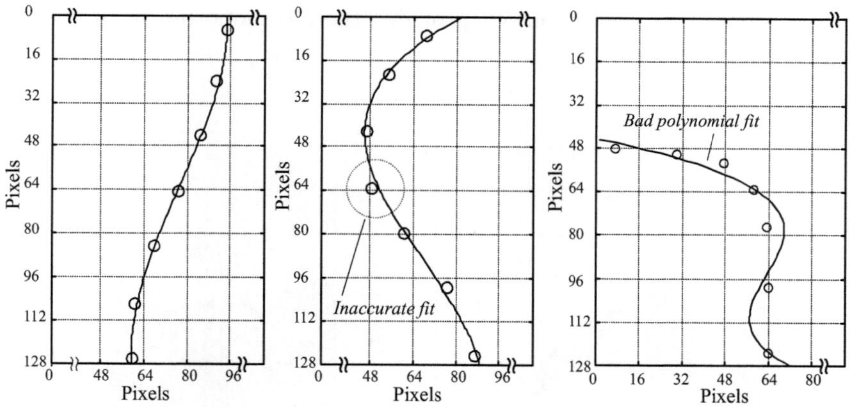

Fig. B.4. Good (**left**), inaccurate (**middle**) and wrong (**right**) 3rd order polynomial fit

For C^2-continuity, the segments need to be cubic (3rd order). Higher order continuity is often inadequate due to the previously mentioned undesirable characteristics of higher order polynomials. We therefore restrict this section to the discussion of cubic splines. Numerous cubic spline representations exist, differing in the way the model parameters are determined and in the extra conditions imposed on the spline. Each cubic spline segment k corresponds to

$$x_k = a_k(y - y_k)^3 + b_k(y - y_k)^2 + c_k(y - y_k) + d_k. \tag{B.8}$$

For *natural cubic splines* the segments are continuous in their first and second derivatives at the knots or

$$s'_{k-1}(y_k) = s'_k(y_k), \ s''_{k-1}(y_k) = s''_k(y_k), \ k = 2, \dots, n-1. \tag{B.9}$$

Choosing the derivatives in the first and the last knot, $s'_1(y_1)$ and $s'_{n-1}(y_n)$, determines all other parameters.

Hermite cubic splines interpolate the knots (x_k, y_k) with a given slope (x'_k, y_k). They are C^1-continuous in the knots. If the slopes (or tangents) in the knots are unknown, they need to be approximated or determined heuristically. H. Späth [78] describes several methods to determine these slopes, one of which aims at an optimally smoothed spline. To this end, the slopes d_k in the knots are determined by a first order approximation of the derivative: $d_k = (x_{k+1} - x_k)/(y_{k+1} - y_k)$, and adjusted to minimize oscillations based on following conditions:

$$\begin{cases} x'_k = 0 & \text{if } d_{k-1}\, d_k < 0 \\ x'_k = x'_{k+1} = 0 & \text{if } d_k = 0 \\ sign(x'_k) = sign(x'_{k+1}) & = sign(d_k). \end{cases} \tag{B.10}$$

This results in a smooth contour provided there is monotonicity in one of the contour coordinates.

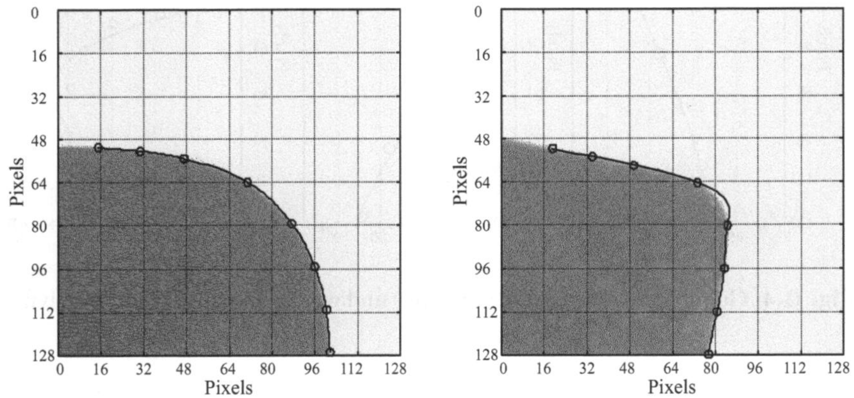

Fig. B.5. Natural cubic splines

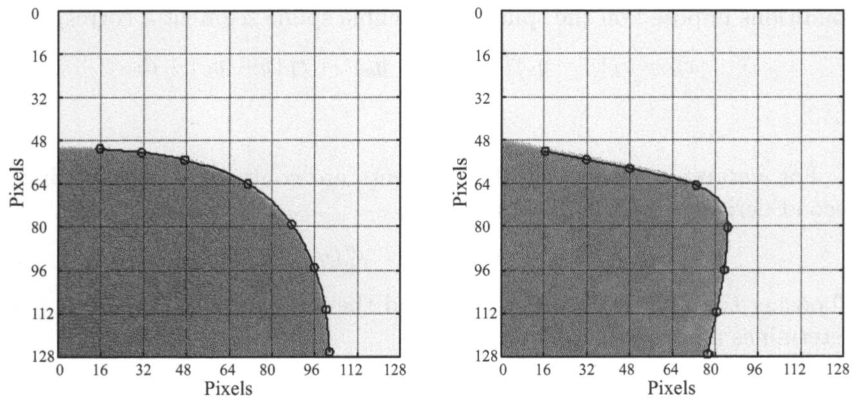

Fig. B.6. Smoothed Hermite cubic splines

Figures B.5 and B.6 give some examples. In general, a cubic spline fit gives a close approximation of the real contour (given the contour points are monotonic), with a slightly better fit for Hermite cubic splines than for natural cubic splines. This automatically results in adequate pose computation. The curvature profile of the spline contour, however, shows unacceptable oscillations as illustrated in Fig. B.7. The computed curvature differs significantly from the known curvature of the circular (testing) contour. This is partly due to the presence of a second derivative in the calculation of the curvature κ:

$$\kappa = \frac{\frac{d^2x}{dy^2}}{\left[1 + \left(\frac{dx}{dy}\right)^2\right]^{\frac{3}{2}}} \tag{B.11}$$

and partly inherent to the very nature of interpolating splines. For reasons of comparison, Fig. B.7 also shows the computed curvature versus arc length resulting from a 3rd order polynomial fit and from the tangent model. Clearly the tangent model gives the best results.

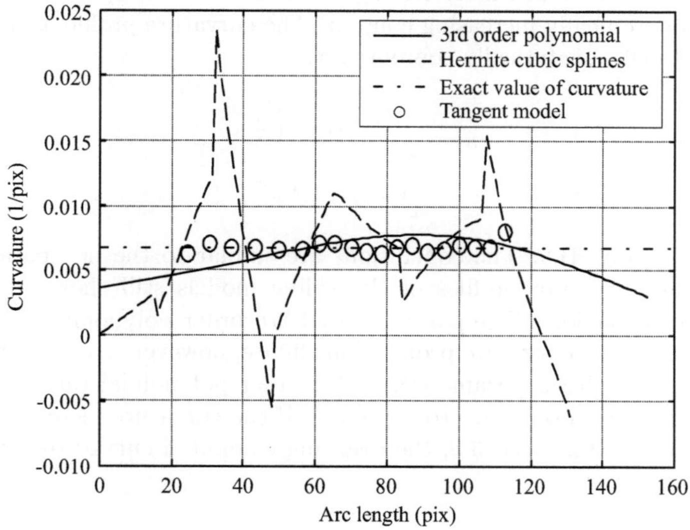

Fig. B.7. Computed curvature versus arc length according to three fitting models for a set of contour points lying on a circle with radius 150 pix; the exact curvature is 0.00667 (1/pix)

B.6 Parameterized Representation

Until now x is expressed as a function of y. For interpolating methods, y then needs to be monotonically increasing (or decreasing) which is not necessarily the case for an arbitrary contour.

The parameterized representation is introduced to relieve the necessity of monotonicity in x or y. In a parameterized representation both x and y are expressed as a function of a monotonically increasing parameter w:

$$x = \xi(w), \quad y = \eta(w). \tag{B.12}$$

The functions ξ and η are determined according to the previously mentioned methods being either non-interpolating 3rd order polynomial or cubic spline.

The simplest choice for w is $w = 0, \ldots, n$-1. If, on the other hand, w is chosen as

$$w_1 = 0, \quad w_{k+1} = w_k + \sqrt{(\Delta x_k)^2 + (\Delta y_k)^2},\ k = 1, \ldots n{-}1, \tag{B.13}$$

then w is a linear approximation of the arc length along the contour and the resulting fit is arc length parameterized.

In general, a parameterized fit approximates the real contour better than a similar non-parameterized fit. This is especially the case for a parameterized 3rd order polynomial. Figure B.8 again gives some examples, this time with a non-monotonically increasing contour. The curvature profile of the parameterized models, analytically computed as

$$\kappa = \frac{\frac{d^2\xi}{dw^2}\frac{d\zeta}{dw} - \frac{d^2\zeta}{dw^2}\frac{d\xi}{dw}}{\left[\left(\frac{d\xi}{dw}\right)^2 + \left(\frac{d\zeta}{dw}\right)^2\right]^{\frac{3}{2}}} \tag{B.14}$$

is shown in Fig. B.9. The conclusions are similar to the non-parameterized cases. The curvature profiles of the spline models still show unacceptable oscillatory behaviour. The parameterized 3rd order polynomial gives a fairly good (computed) curvature profile. This figure, however, gives the best result ever achieved with a parameterized 3rd order polynomial (and is only that good for the given set of contour points). If the curvature increases (as is the case in Fig. 3.32 in Sect. 3.8, the resulting computed curvature profile is not that good.

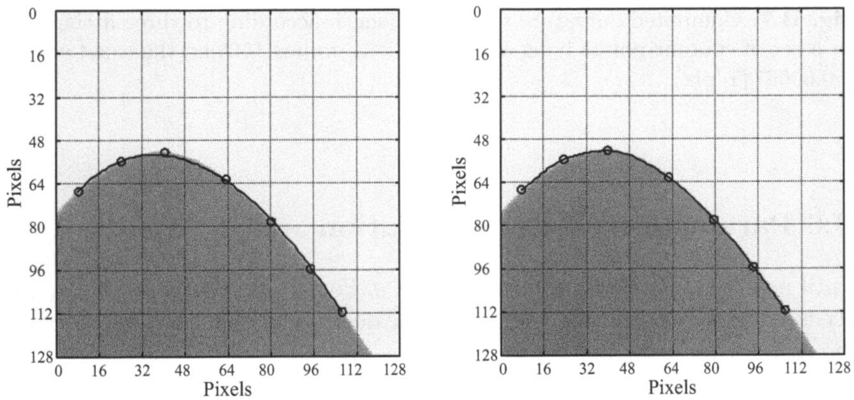

Fig. B.8. Examples of parameterized fits for a non-monotonically increasing contour; (**left**) parameterized 3rd order polynomial; (**right**) parameterized Hermite cubic splines

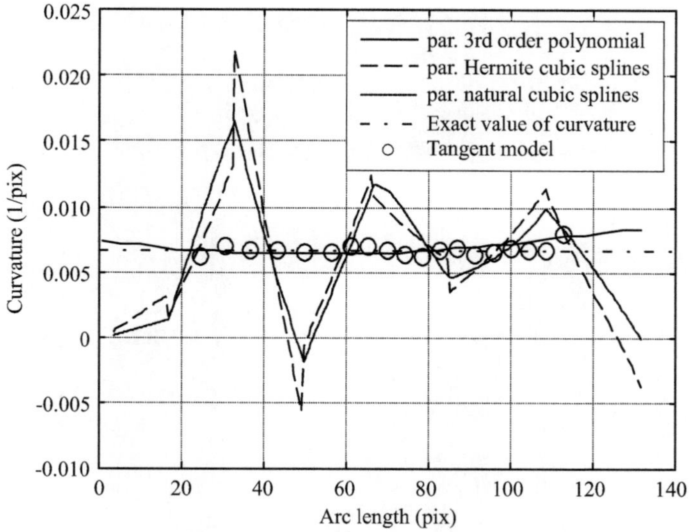

Fig. B.9. Computed curvature versus arc length according to four fitting models for a set of contour points lying on a circle with radius 150 pix; the exact curvature is 0.00667 (1/pix)

B.7 Conclusion

The most important criteria in the comparison of the different proposed contour fitting models is clearly the accuracy of the contour curvature computation. Only the tangent model gives satisfying computed curvature values. Clearly, the twice applied, once for the tangent and once for the curvature, least squares solution in the tangent model case gives better results than any other approach. Most of the proposed models have a good positional accuracy. However, as several examples illustrate, this is no guarantee whatsoever for a good curvature computation.

Section 3.8 already compared the computational implications of the three best contour models: the tangent model, the parameterized 3rd order polynomial and the parameterized Hermite splines. Again, the tangent model is the most efficient one.

The tangent model only uses local contour information. This proves to be important for a good curvature computation, even in the case of a constant curvature profile. Other contour models too, e.g. a parameterized second order polynomial, can be applied locally. Unfortunately however, they still imply a bigger computational effort than the tangent model and do not give better computed control data.

As well known, an alternative to the least squares implementation is the use of a KALMAN filter for the estimation of the model and/or control parameters. Both methods will give the same results. A KALMAN filter then provides the option to take into account the measurement noise in the parame-

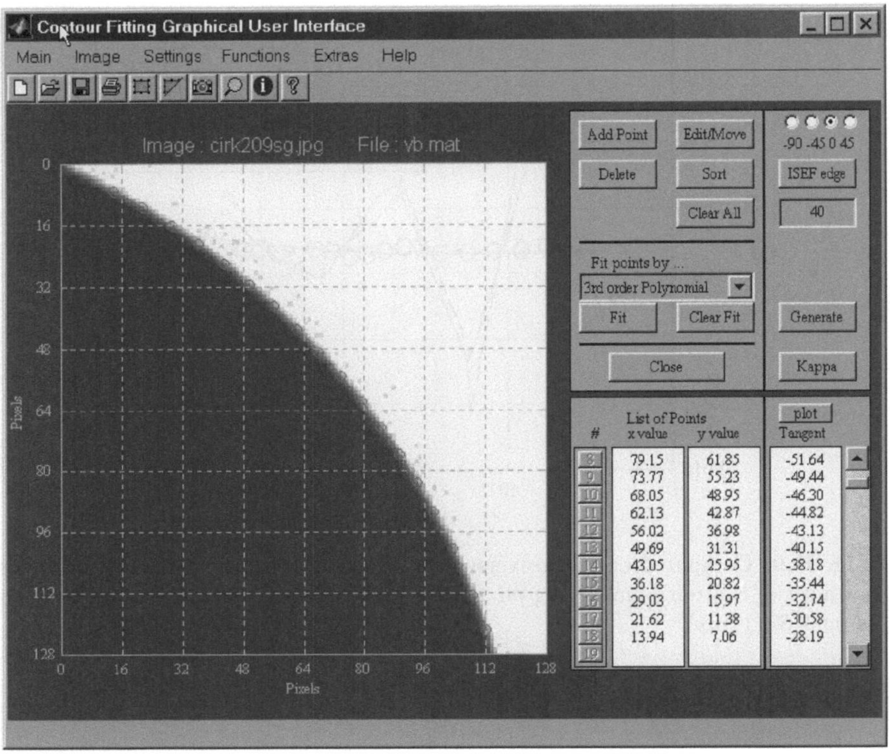

Fig. B.10. Main window of the Contour Fitting Graphical User Interface

ter estimation. Given the satisfying results of the tangent model, however, this option is not pertinent and hence not further investigated in our approach.

B.8 Graphical User Interface

All the presented models are implemented in MATLAB®. The Contour Fitting Graphical User Interface (CFGUI) gives a user friendly and interactive environment to test and compare the contour fitting models. Figure B.10 shows the layout of the CFGUI. Its functionality includes, among others,

- file management for images and contour data;
- support of PGM, GRY and JPG (8-bit grayscale) images with size up to 512 by 512 pixels;
- on line image grabbing (from DSP);
- interactive buildup of up to 200 contour points, graphically as well as numerically (add, delete, move, sort)
- interactive as well as automatic contour extraction by ISEF edge detection under four direction;

- up to 13 different contour fitting models;
- tangent and curvature computation of the fitted contour;
- postscript print out;
- automatic path generation for KUKA361 robot from the fitted contour.

C

Experimental Validation of Curvature Computation

This appendix investigates experimental results for the curvature computation on a constant curved contour.

A correctly computed contour curvature κ is essential for the success of the curvature based feedforward control (according to (6.18)). In order to compare the computed curvature with the correct theoretical curvature, a piecewise constant curved test object is used. The dimensions of the test object are given in Sect. E.2. See also Fig. C.3.

Figure C.1 shows measured and actual curvature versus arc length with a forward velocity of 40 mm/s. If the curvature is constant, the computed value corresponds well with the actual one. At the step transitions, the computed curvature shows a more smoothed profile. In view of the used computation method, this is only logical. After all, the computed curvature is the mean curvature over a finite contour segment. With an increasing number of contour points used in the curvature computation or with an increasing tangent velocity, the length of this finite contour segment also increases. The longer the contour segment, the more a transient in the curvature profile is smoothed. On the other hand, the noise sensitivity decreases if the contour segment over which the mean curvature is calculated, increases. Figures C.1 and C.2 give the experimental verification.

Figure C.1 shows the computed curvature on the basis of 15, 11, 9 and 5 contour points. Less points gives a more noisy signal, which however follows the transitions a little better. For five points, the computed curvature becomes too noisy and hence useless.

Figure C.2 shows the computed curvature for an increased velocity of 45, 70 and 110 mm/s. For these velocities, only the bigger arcs are suited to be followed. At the highest velocity, the actual step transition is smoothed in such a way that the resulting force profile[1] with feedforward control is just barely acceptable.

[1] not shown here

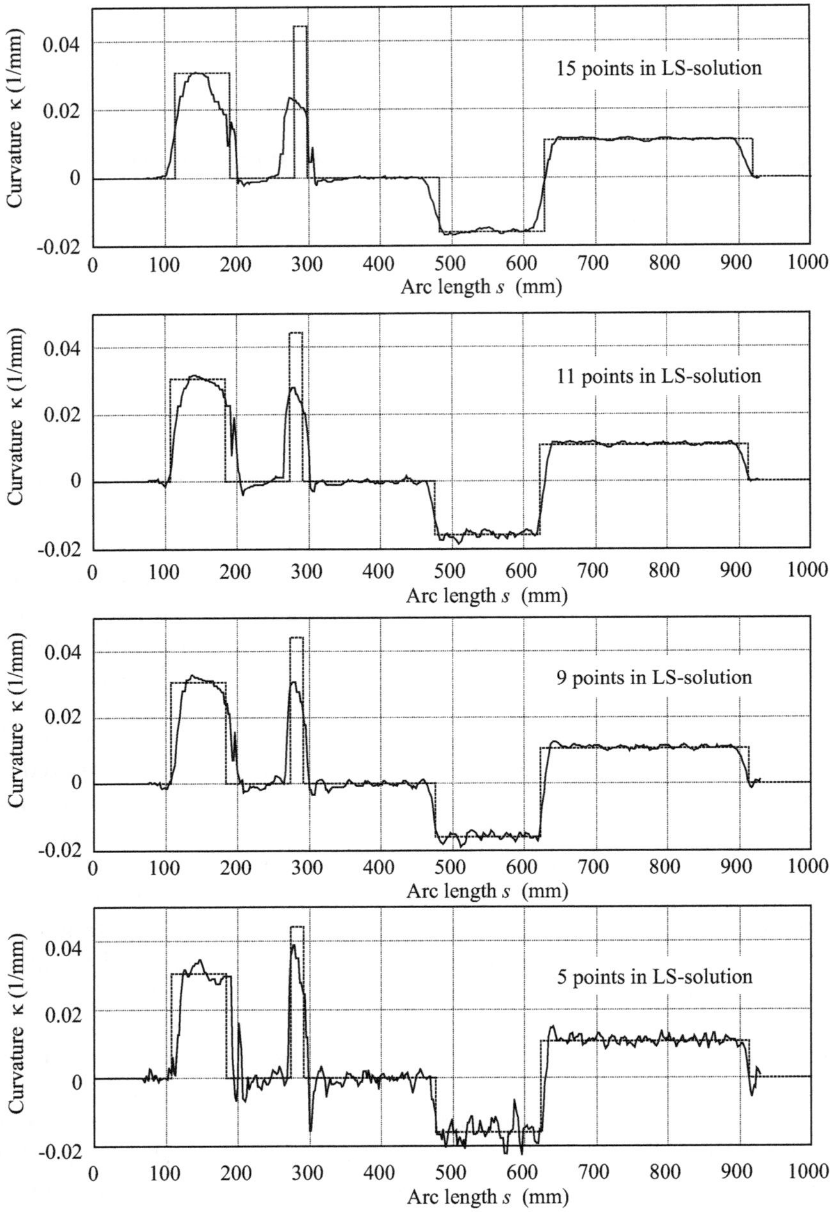

Fig. C.1. Measured and actual curvature of a constant curved object with velocity 40 mm/s using 15, 11, 9 or 5 contour points in the curvature computation respectively

For the sake of completeness, Fig. C.3 shows the corrected offset contour of the constant curved object as measured by the vision system and Fig. C.4 gives the measured contour orientation versus arc length. This figure clearly

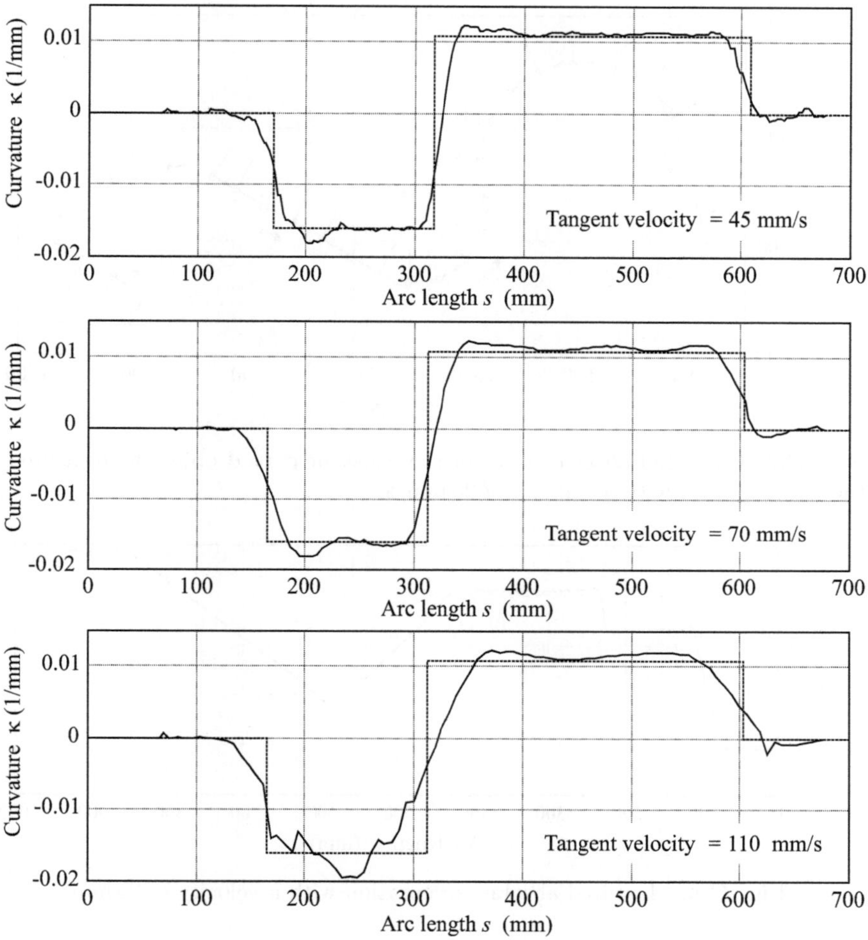

Fig. C.2. Measured curvature of a constant curved object with velocities set to 45, 70 and 110 mm/s

indicates the fact that the curvature is no more than the tangent of the shown orientation profile. A numerical differentiation of this profile would however result in a far too noisy curvature signal.

Other experimental results, e.g. for a sinusoidal contour, are given by Verbiest and Verdonck in [87].

The shown experimental results validate the used method for the curvature computation. At constant curved segments the computed curvature equals the actual curvature. At (step) transitions in the actual curvature, the measured curvature shows a much more smoothed profile. Here, a trade-off between accuracy and noise sensitivity has to be made, by choosing the optimal number

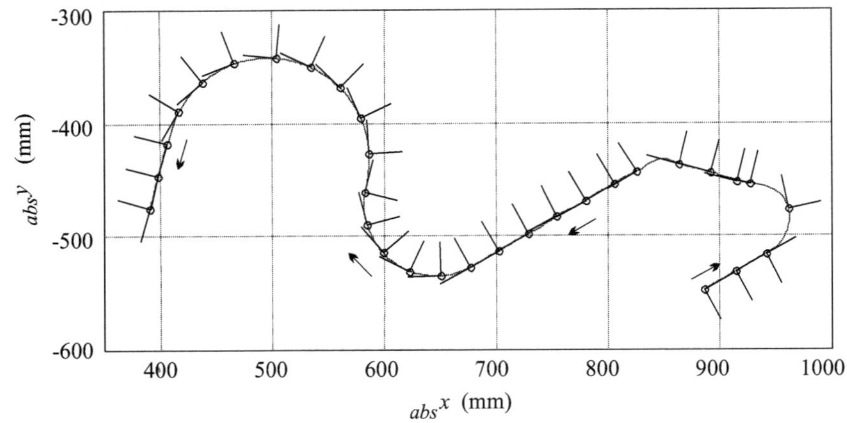

Fig. C.3. Corrected offset contour of the constant curved object as measured by the vision system with a velocity of 25 mm/s

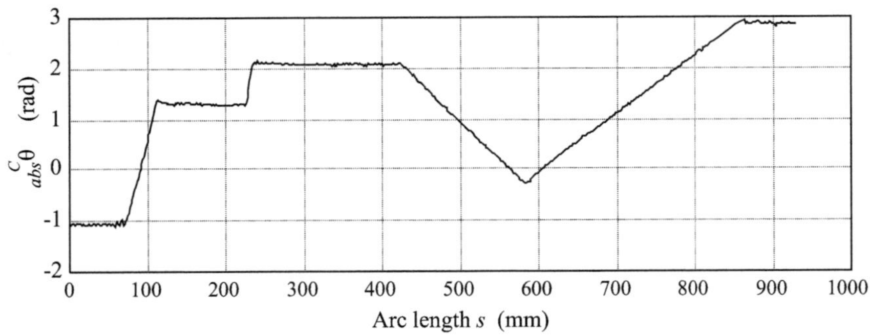

Fig. C.4. Measured absolute orientation with a velocity of 25 mm/s

of contour points (or the optimal length of the contour segment) to be used in the least squares curvature computation. The experimental results give an optimum which lies between 9 and 11 points.

The accuracy of the curvature computation, can improve if more contour points, lying more closely together, are measured and used. This involves the detection of more than one contour point per image[2], hereby introducing however new problems in ordering the detected contour points from image to image and pushing the computation time to its limits.

[2] with unchanged video frame rate

D

Image Processing Implementation on DSP

This appendix describes the basic elements of the implemented image processing software on the DSP-C40 (digital signal processor).

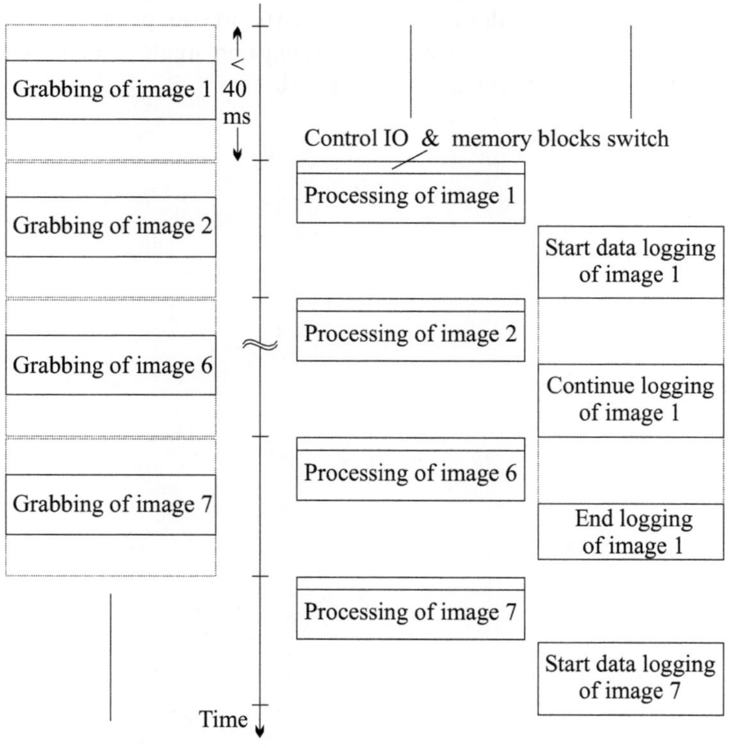

Fig. D.1. Time chart of main program loop relating grabbing, processing and logging processes

The image processing software is mainly written in C. Some libraries are originally written in Assembler, in order to optimize the processing time. The image processing executables are built on top of the real-time multi-processor operating platform, VIRTUOSO, based 1) on the image grabbing libraries for the DSP/VSP configuration provided by HEMA, 2) on the C-libraries and compiler for the DSP-C40 of Texas Instruments, 3) on the Meschach C-library to compute the least squares solutions and 4) on the own implemented main program and routines, e.g. to compute the relative area parameters or to detect the ISEF edge.

The DSP/VSP unit can fully simultaneously grab a new image and process a previously grabbed image. Three memory blocks are used to store 1) the newly grabbed image, 2) the currently processed image and 3) the currently saved image. Figure D.1 gives the basic time chart of a standard program.

Typically, without IO to the host, the processing of one image, consisting of unpacking the image, contour extraction for 9 to 11 points, least squares line fit, position match and least squares curvature fit, takes about 10 to 15 ms. The remaining time is used for IO with the host to display messages, log data and save images. The former process is avoided as much as possible. The latter are programmed as parallel processes with the lowest priority , in order not to disturb the real time control cycle. Saving one single image spans several cycles. Hence, not every image can be saved.

E

Technical Documentation

E.1 Sinusoidal Contour

The first testing object is a 30 mm tick PVC-plate. One of it sides has a sinus-like profile as shown in Fig. E.1. The trajectory of the milling tool, with radius (12.5 mm), is a perfect sine. This means that the 'peaks' (curves 1 and 3) of test object are curved more sharply than the 'valley' (curve 2). The sine has an amplitude of 75 mm and a period of 300 mm. The colour of the test object is dark grey, making it clearly distinguishable from a bright surrounding.

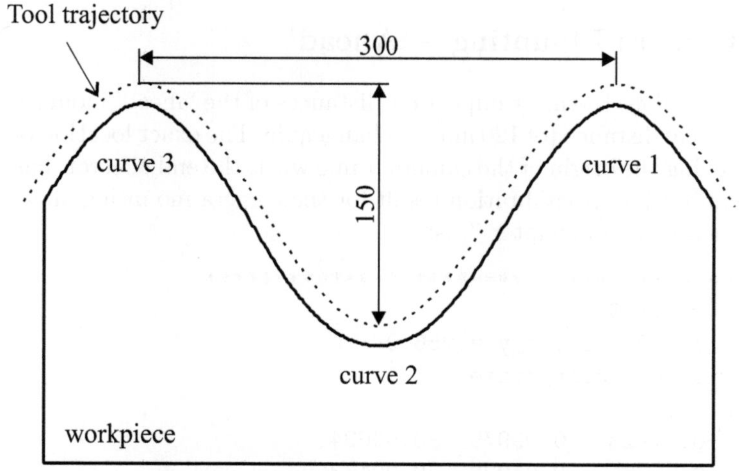

Fig. E.1. Test object with sinusoidal contour, scale 1:4

E.2 'Constant-Curved' Contour

The second test object has a piecewise constant curved contour. It thus consists of straight lines and arc shaped segments. Using a constant curved contour allows a verification of the correctness of the feedforward calculation. To give a good contrast with the environment, the top layer of the object is painted black. Figure E.2 shows the exact dimensions of the this object.

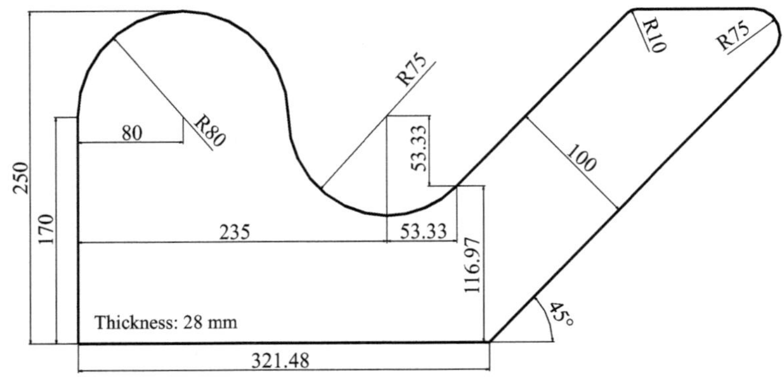

Fig. E.2. Piecewise constant curved test object, scale 1:6

E.3 Camera Mounting – 'Ahead'

Figure E.3 gives the most important distances of the 'ahead' mounted camera. The vertical distance (of 120 mm) is changeable. The exact location of the focal point, being the origin of the camera frame w.r.t. the end effector, has to follow from calibration. A calibration result for the camera mounting ahead, used in the experiments of chapter 7, is:

```
*************************************
f = 6.146
c_x = 255.9 ; c_y = 256.0
vision_sensor_frame :
R = [
 -0.01923   0.99979  -0.00634
  0.99981   0.01928   0.00173
  0.00185  -0.00630  -0.99997 ]
T = [
 0.51178e+002
-0.01877e+002
 2.18613e+002 ]
*************************************
```

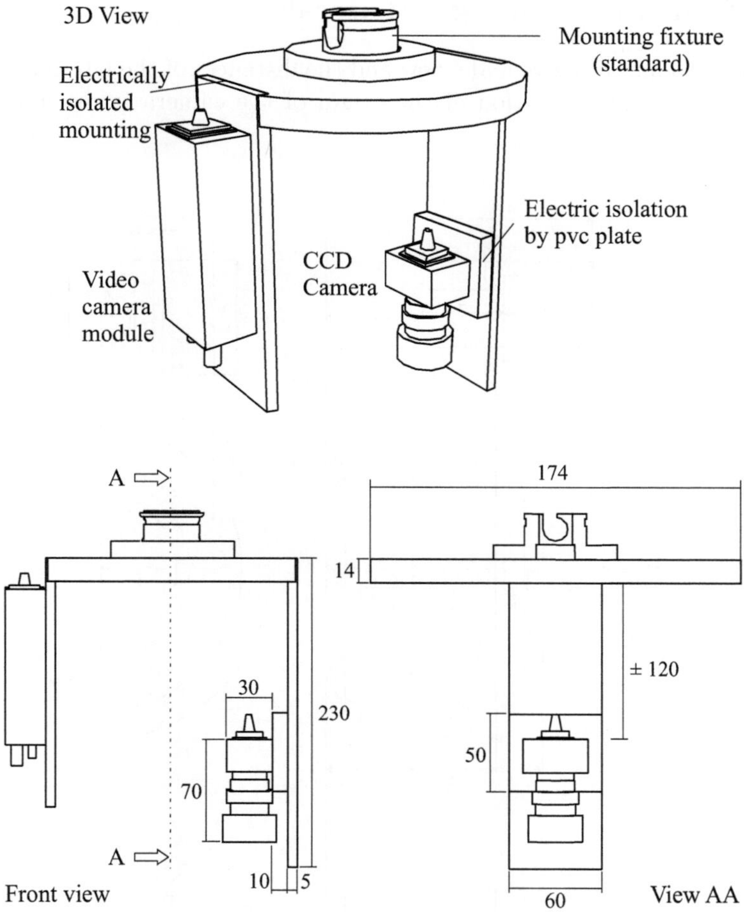

Fig. E.3. Camera mounting 'ahead'

E.4 Camera Mounting – 'Lateral'

Figures E.4 and E.5 give a 3D view and the distances of the laterally mounted camera. The exact location of the origin of the camera frame follows from calibration.

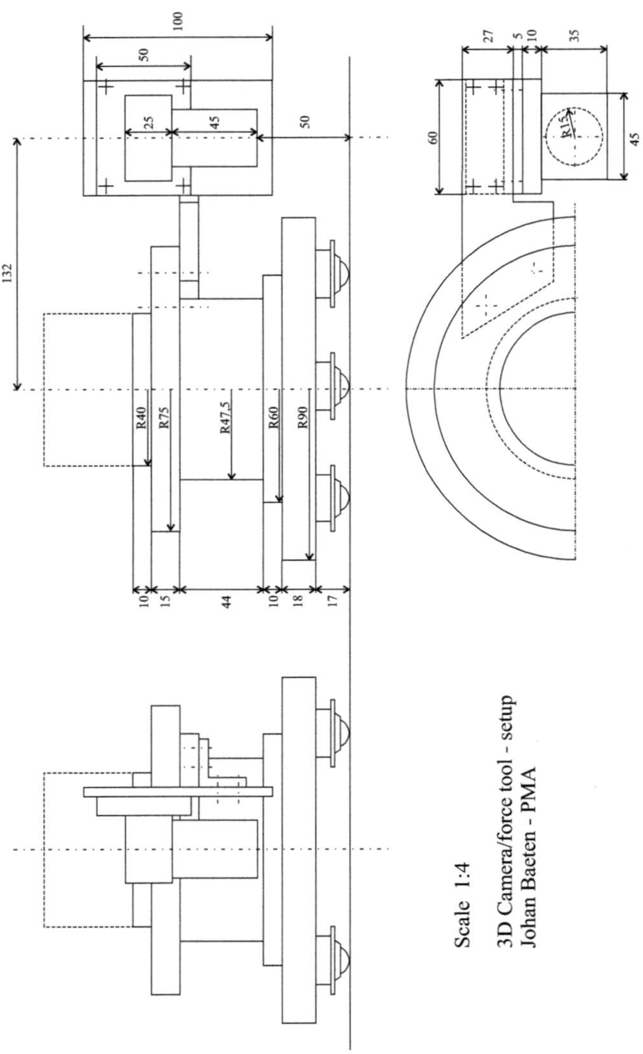

Fig. E.4. Technical drawing of lateral camera mounting

A calibration result for the lateral camera mounting, used in the third experiment of chapter 8, is

```
*************************************
f = 6.192
c_x = 255.4  ; c_y =256.9
vision_sensor_frame :
R = [
 -4.558e-002  9.989e-001   6.706e-003
  9.987e-001  4.571e-002  -2.079e-002
 -2.108e-002  5.750e-003  -9.998e-001 ]
T = [
  6.500e-002
 -1.467e+002
  4.560e+001]
*************************************
```

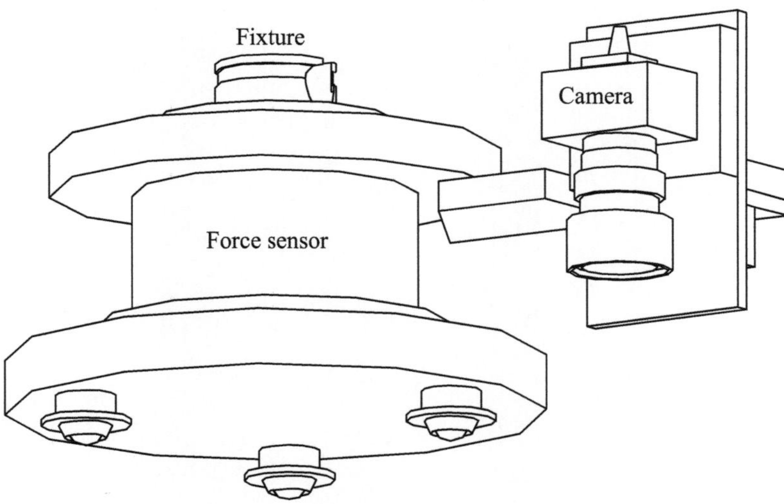

Fig. E.5. 3D view of lateral camera mounting

References

1. Peter K. Allen, Aleksandar Timcenko, Billibon Yoshimi, and Paul Michelman. Automated tracking and grasping of a moving object with a robotic hand-eye sytem. *IEEE Trans. on Robotics and Automation*, 9(2):152–165, April 1993.
2. Minoru Asada, Takamaro Tanake, and Koh Hosoda. Visual tracking of unknown moving object by adaptive binocular visual servoing. In *Int. Conf. on Multisensor Fusion and Integration for Intel. Syst.*, pages 249–254, Taipai, Taiwan, August 1999.
3. Johan Baeten, Herman Bruyninckx, and Joris De Schutter. Combining eye-in-hand visual servoing and force control in robotic tasks using the task frame. In *Int. Conf. on Multisensor Fusion and Integration for Intel. Syst.*, pages 141–146, Taipai, Taiwan, August 1999.
4. Johan Baeten, Herman Bruyninckx, and Joris De Schutter. Shared control in hybrid vision/force robotic servoing using the task frame. In *Int. Conf. on Intelligent Robots and Systems*, pages 2128–2133, EPFL, Lausanne, Switzerland, May 2002.
5. Johan Baeten, Herman Bruyninckx, and Joris De Schutter. Tool/camera configurations for eye-in-hand hybrid vision/force control. In *IEEE Int. Conf. on Robotics and Automation*, pages 1704–1709, Washington, DC, May 2002.
6. Johan Baeten and Joris De Schutter. Improving force controlled planar contour following using on-line eye-in-hand vision based feedforward. In *Proc. of Int. Conf. on Advanced Intelligent Mechatronics*, pages 902–907, Atlanta, GA, September 1999.
7. Johan Baeten and Joris De Schutter. Combined vision/force control at corners in planar robotic contour following. In *Proc. of Int. Conf. on Advanced Intelligent Mechatronics*, pages 810–815, Como, Italy, July 2001.
8. Johan Baeten and Joris De Schutter. Hybrid vision/force control at corners in planar robotic contour following. *IEEE/ASME Transactions on Mechatronics*, 7(2):143–151, June 2002.
9. Johan Baeten, Walter Verdonck, Herman Bruyninckx, and Joris De Schutter. Combining force control and visual servoing for planar contour following. *Machine Intelligence and Robotic Control*, 2(2):69–75, July 2000.
10. Herman Bruyninckx and Joris De Schutter. Specification of force-controlled actions in the "task frame formalism" – A synthesis. *IEEE Trans. on Robotics and Automation*, 12(4):581– 589, August 1996.

11. J.F. Canny. Finding edges and lines in images. Master's thesis, MIT, Cambridge, USA, 1983.
12. J.F. Canny. A computational approach to edge detection. *IEEE Trans. on Pattern Analysis and Machine Intelligence*, 8(6):679–698, November 1986.
13. C. Canudas De Wit et al. *Theory of Robot Control.* Springer, London, 1996. pp. 150-170.
14. François Chaumette, Patrick Rives, and Bernard Espiau. Positioning of a robot with respect to an object, tracking it and estimating its velocity by visual servoing. In *IEEE Int. Conf. on Robotics and Automation*, pages 2248–2253, Sacramento, CA, 1991.
15. Gil Whoan Chu and Myung Jin Chung. Selection of an optimal camera position using visibility and manipulability measures for an active camera system. In *Int. Conf. on Intelligent Robots and Systems*, pages 429–434, Takamatsu, Japan, 2000.
16. Christophe Collewet and François Chaumette. A contour approach for image-based control on objects with complex shape. In *Int. Conf. on Intelligent Robots and Systems*, pages 751–756, Takamatsu, Japan, 2000.
17. Peter I. Corke and M. Good. Dynamic effects in high-performance visual servoing. In *IEEE Int. Conf. on Robotics and Automation*, pages 1838–1843, Nice, France, 1992.
18. Peter I. Corke. Dynamics of visual control. In *Workshop M-5, Visual Servoing: Achievements, Applications and Open Problems, IEEE International Conference on Robotics and Automation*, 1994. San Diego, California.
19. Peter I. Corke. Visual control of robot manipulators – A review. In *Workshop M-5, Visual Servoing: Achievements, Applications and Open Problems, IEEE International Conference on Robotics and Automation*, pages 1–31, 1994. San Diego, California.
20. Peter I. Corke. *Dynamic issues in robot visual-servo systems*, pages 488–498. G. Giralt and G. Hirzinger, editors, Robotics Research. The Seventh International Symposium. Springer-Verlag, 1996.
21. Peter I. Corke. *Visual Control of Robots: High-Performance visual servoing.* Mechatronics. Research Studies Press (John Wiley), 1996.
22. Eve Coste-Manière, Phillipe Couvignou, and Pradeep K. Khosla. Contour following based on visual servoing. In *Int. Conf. on Intelligent Robots and Systems*, pages 716–722, Yokohama, Japan, July 1993.
23. Eve Coste-Manière, Phillipe Couvignou, and Pradeep K. Khosla. Visual servoing in the task-function framework: a contour following task. *Journal of Intelligent and Robotic Systems*, 12:1–21, 1995.
24. Sabine Demey, Herman Bruyninckx, and Joris De Schutter. Model-based planar contour following in the presence of pose and model errors. *Int. J. Robotics Research*, 16(6):840–858, 1996.
25. Sabine Demey, Stafan Dutré, Wim Persoons, P. Van De Poel, Wim Witvrouw, Joris De Schutter, and Hendrik Van Brussel. Model based and sensor based programming of compliant motion tasks. In *Proc. Int. Symposium on Industrial Robots*, pages 393–400, 1993. Tokyo.
26. Joris De Schutter, Herman Bruyninckx, Sabine Demey, and Wim Witvrouw. Force controlled robots tutorial. In *Technology Transfer Workshop on Industrial Vision and Autonomous Robots*, June 1994. Leuven, Belgium.

27. Joris De Schutter and Hendrik Van Brussel. Compliant robot motion. II. A control approach based on external control loops. *Int. J. Robotics Research*, 7(4):18–33, 1988.
28. Joris De Schutter and Hendrik Van Brussel. Compliant robot motion. I. A formalism for specifying compliant motion tasks. *Int. J. Robotics Research*, 7(4):3–17, 1988.
29. Joris De Schutter, P. Van de Poel, Wim Witvrouw, Herman Bruyninckx, Sabine Demey, and Stefan Dutré. An environment for experimental compliant motion. In *Tutorial on Force and Contact Control in Robotic Systems: A historical Perspective and Current Technologies, IEEE Conf. on Robotics and Automation*, pages 112–126, 1993. Atlanta.
30. Joris De Schutter. Compliant robot motion: Task formulation and control, 1986. Ph.D. dissertation, Departement of Mechanical Engineering, Katholieke Universiteit Leuven, Belgium.
31. Joris De Schutter. Improved force control laws for advanced tracking applications. In *IEEE Int. Conf. on Robotics and Automation*, pages 1497–1502, Philadelphia, PA, 1988.
32. T. Drummond and R. Cipolla. Real-time tracking of complex structures with on-line camera calibration. In *Britisch Machine Vision Conference*, 1999.
33. T. Ellis, A. Abbood, and B. Brillaullt. Ellipse detection an matching with uncertainty. *Image and Vision Computing*, 10(5):271–276, June 1992.
34. Bernard Espiau, Chaumette François, and Patrick Rives. A new approach to visual servoing in robotics. *IEEE Trans. on Robotics and Automation*, 8(3):313–326, June 1992.
35. John T. Feddema and Owen R. Mitchell. Vision-guided servoing with feature-based trajectory generation. *IEEE Trans. on Robotics and Automation*, 5(5):691–700, October 1989.
36. G. Hager and S. Hutchinson. *Visual Servoing: Achievements, Applications and Open Problems.* Workshop M-5 of IEEE Int. Conf. on Robotics and Automation, San Diego, CA, 1994.
37. R. S. Hartenberg and J. Denavit. A kinematic notation for lower pair mechanisms based on matrices. *Journal of Applied Mechanics*, 77:215–221, June 1955.
38. Koichi Hashimoto, Tsutomu Kimoto, Takumi Ebine, and Hidenori Kimura. Manipulator control with image-based visual servo. In *IEEE Int. Conf. on Robotics and Automation*, pages 2267–2271, Sacramento, CA, 1991.
39. G. Hirzinger. Direct digital control using a force-torque sensor. In *Proc. IFAC Symp. on Real Time Digital Control Applications*, 1983.
40. N. Hogan. Impedance control: An approach to manipulation: Part i-theory; part ii-implementation; part iii-applications. *ASME J. Dynamic Systems, Measurements and Control*, 107(1):1–24, 1985.
41. B. Horn and B. Schunck. Determining optical flow. *Artificial Intelligence*, 17:185–203, 1981.
42. Koh Hosoda and Minoru Asada. Versatile visual servoing without knowledge of true jacobian. In *Int. Conf. on Intelligent Robots and Systems*, pages 186–193, München, Germany, 1994.
43. Koh Hosoda, Katsuji Igarashi, and Minoru Asada. Adaptive hybrid control for visual and force servoing in an unknown environment. *IEEE Rob. Automation Mag.*, 5(4):39–43, December 1998.
44. S. Hutchinson, G.D. Hager, and P.I. Corke. A tutorial on visual servo control. *IEEE Trans. on Robotics and Automation*, 12(5):651–670, October 1996.

45. Michael Isard and Andrew Blake. Contour tracking by stochastic propagation of conditional density. In *Proceedings of European Conference on Computer Vision*, pages 343–356, 1996.
46. Idaku Ishii, Yoshihiro Nakabo, and Masatoshi Ishikawa. Target tracking algorithm for 1ms visual feedback system using massively parallel processing. In *Proceedings of IEEE International Conference on Robotics and Automation*, pages 2309–2314, 1996. Minneapolis, Minnesota.
47. Stefan Jörg, Jörg Langwald, Johannes Stelter, Gerd Hirzinger, and Ciro Natale. Flexible robot-assembly using a multi-sensory approach. In *IEEE Int. Conf. on Robotics and Automation*, pages 3687–3694, San Francisco, CA, 2000.
48. Ho Won Kim, Jae Seung Cho, and In So Kweon. A novel image-based control-law for the visual servoing system under large pose error. In *Int. Conf. on Intelligent Robots and Systems*, pages 263–268, Takamatsu, Japan, 2000.
49. Friedrich Lange and Gerd Hirzinger. A universal sensor control architecture considering robot dynamics. In *Int. Conf. on Multisensor Fusion and Integration for Intel. Syst.*, pages 277–282, Germany, 2001.
50. Friedrich Lange, Patrick Wunsch, and Gerard Hirzinger. Predictive vision based control of high speed industrial robot paths. In *IEEE Int. Conf. on Robotics and Automation*, pages 2646–2651, Leuven, Belgium, 1998.
51. Sung-Woo Lee, Burn-Jae You, and Gregory D. Hager. Model-based 3-d object tracking using projective invariance. In *IEEE Int. Conf. on Robotics and Automation*, pages 1598–1594, Detroit, MI, 1999.
52. R.K. Lenz and R.Y. Tsai. Techniques for calibration of the scale factor and image center for high accuracy 3d machine vision metrology. In *IEEE Int. Conf. on Robotics and Automation*, pages 68–75, Raleigh, NC, 1987.
53. Ezio Malis, Guillaume Morel, and François Chaumette. Robot control using disparate multiple sensors. *Int. J. Robotics Research*, 20(5):364–377, May 2001.
54. Eric Marchand. Visp: A software environment for eye-in-hand visual servoing. In *IEEE Int. Conf. on Robotics and Automation*, pages 3224–3229, Detroit, MI, 1999.
55. M.T. Mason. Compliance and force control for computer controlled manipulators. *IEEE Trans. on Systems, Man, and Cybernetics*, 11:418–432, 1981.
56. Guillaume Morel, Ezio Malis, and Sylvie Boudet. Impedance based combination of visual and force control. In *IEEE Int. Conf. on Robotics and Automation*, pages 1743–1748, Leuven, Belgium, 1998.
57. Keiro Nagahama, Koichi Hashimoto, Toshiro Noritsugu, and Masahiro Takaiwa. Visual servoing based on object motion estimation. In *Int. Conf. on Intelligent Robots and Systems*, pages 245–250, Takamatsu, Japan, 2000.
58. Yoshihiro Nakabo and Masatoshi Ishikawa. Visual impedance using a 1 ms visual feedback system. In *IEEE Int. Conf. on Robotics and Automation*, pages 2333–2338, Leuven, Belgium, 1998.
59. Akio Namiki, Yoshihiro Nakabo, Idaku Ishii, and Masatoshi Ishikawa. High speed grasping using visual and force feedback. In *IEEE Int. Conf. on Robotics and Automation*, pages 3195–3200, Detroit, MI, 1999.
60. Bradley J. Nelson and Pradeep K. Khosla. Force and vision resolvability for assimilating disparate sensory feedback. *IEEE Trans. on Robotics and Automation*, 12(5):714–731, October 1996.
61. Bradley J Nelson, J.D. Morrow, and Pradeep K. Khosla. Robotic manipulation using high bandwidth force and vision feedback. *Mathl. Computer Modelling*, 24(5/6):11–29, 1996.

62. Bradley J. Nelson, Nikolaos P. Papanikolopoulos, and Pradeep K. Khosla. Robotic visual servoing and robotic assembly tasks. *IEEE Rob. Automation Mag.*, pages 23–31, June 1996.
63. Bradley J. Nelson, Yu Zhou, and Barmeshwar Vikramaditya. Integrating force and vision feedback for microassembly. In *Proc. of SPIE: Microrobotics and Microsystems Fabrication*, pages 30–41, October 1997. Pittsburgh, PA.
64. Bradley J. Nelson. Assimilating disparate sensory feedback within virtual environments for telerobotic systems. *Robotics and Autonomous Systems*, 36:1–10, 2001.
65. Nikolaos P. Papanikolopoulos, Pradeep K. Khosla, and Takeo Kanade. Visual tracking of a moving target by a camera mounted on a robot: A combination of control and vision. *IEEE Trans. on Robotics and Automation*, 9(1):14–35, February 1993.
66. Nikolaos P. Papanikolopoulos and Pradeep K. Khosla. Feature based robotic visual tracking of 3-d translational motion. In *Proc. of 30th IEEE conf. on Decision and Control*, December 1991. Brighton, United Kingdom.
67. Andreas Pichler and Martin Jägersand. Uncalibrated hybrid force-vision manipulation. In *Int. Conf. on Intelligent Robots and Systems*, pages 1866–1871, Takamatsu, Japan, 2000.
68. Jenelle Armstrong Piepmeier, Gary V. McMurray, and Harvey Lipkin. A dynamic jacobian estimation method for uncalibrated visual servoing. In *Proc. of Int. Conf. on Advanced Intelligent Mechatronics*, pages 944–949, Atlanta, GA, 1999.
69. M. Raibert and J. J. Craig. Hybrid position/force control of manipulators. *Trans. ASME J. Dyn. Systems Meas. Control*, 102:126–133, 1981.
70. Arthur C. Sanderson and Lee E. Weiss. *Adaptive Visual Servo Control of Robots*, pages 107–116. A. Pugh, editor, Robot Vision, Int. Trends in Manufactoring Technology. Springer-Verlag, 1983.
71. M. Sato and K.I. Aggerwal. Estimation of position and orientation from image sequence of a circle. In *IEEE Int. Conf. on Robotics and Automation*, pages 2252–2257, Albuquerque, NM, 1997.
72. Peter Schillebeeckx and Swen Vandenberk. Contourvolgen van een robot door het combineren van kracht- en visiecontrole, 1998.
73. J. Shen and S. Castan. An optimal linear operator for step edge detection. *Graphical Models and Image Processing*, 54(2):112–133, March 1992.
74. Bruno Siciliano and Luigi Villani. *Robot Force Control.* The Kluwer Int. series in engineering and computer science. Kluwer Academic Publishers, January 2000. ISBN 0-7923-7733-8.
75. Christopher E. Smith, Scott A. Brandt, and Nikolaos P. Papanikolopoulos. Eye-in-hand robotic tasks in uncalibrated environments. *IEEE Trans. on Robotics and Automation*, 13(6):903–914, December 1997.
76. S.M. Smith and J.M. Brady. Susan - a new approach to low level image processing. *Int. J. Computer Vision*, 23(1):45–78, May 1997.
77. You Song, Wang Tianmiao, Wei Jun, Yang Fenglei, and Zhang Qixian. Shared control in intelligent arm/hand teleoperated system. In *IEEE Int. Conf. on Robotics and Automation*, pages 2489–2494, Detroit, MI, 1999.
78. H. Späth. *One Dimensial Spline Interpolation Algorithms*, pages 151–153. A.K.Peters Ltd., Wellesley, Massachusetts, 1995.

79. M. Tonko, K. Schäfer, F. Haimes, and H.-H. Nagel. Towards visually servoed manipulation of car engine parts. In *IEEE Int. Conf. on Robotics and Automation*, pages 3166–3171, Albuquerque, NM, 1997.
80. R.Y. Tsai and R.Y. Lenz. A new technique for fully autonomous and efficient 3d robotics hand-eye calibration. *IEEE Trans. on Robotics and Automation*, pages 345–358, June 1989.
81. R.Y. Tsai. A versatile camera calibration technique for high-accuracy 3d machine vision metrology using off-the-shelf tv cameras and lenses. *IEEE J. Rob. Automation*, 3(4):323–344, 1987.
82. Tine Tuytelaars, Luc Van Gool, L. D'haeneand, and R. Koch. Matching of affinely invariant regions for visual servoing. In *IEEE Int. Conf. on Robotics and Automation*, pages 1601–1606, Detroit, MI, 1999.
83. P. Van De Poel, Wim Witvrouw, Herman Bruyninckx, and Joris De Schutter. An environment for developing and optimising compliant robot motion tasks. In *6th Int. Conference on Advanced Robotics*, pages 713–718, 1993. Tokyo.
84. Luc Van Gool and M. Proesmans. *Machine Vision*. IMPRO Course 2.2. KU-Leuven, 1995.
85. S. Van Huffel and J. Vandewalle. *The total least squares problem: computational aspects and analysis*. SIAM Philadelphia (Pa.), 1991.
86. S. Venkatesha and R. Owens. An energy feature detection scheme. In *Proceedings of IEEE Int. Conf. on Image Processing*, pages 553–557, September 1989. Singapore.
87. Jan Verbiest and Walter Verdonck. Gebruik van visie voor het verbeteren van krachtgecontroleerd contourvolgen met een robot, 1999. Masters thesis 99EP4, Katholieke Universiteit Leuven, Dept. Werktuigkunde, Afd. PMA.
88. York von Collani, Markus Ferch, Jianwei Zhang, and Alois Knoll. A general learning approach to multisensor based control using statistic indices. In *IEEE Int. Conf. on Robotics and Automation*, pages 3221–3226, San Francisco, CA, 2000.
89. York von Collani, Christian Scheering, Jianwei Zhang, and Alois Knoll. A neuro-fuzzy solution for integrated visaul and force control. In *IEEE Int. Conf. on Robotics and Automation*, pages 2965–2970, Detroit, MI, 1999.
90. York von Collani, Jianwei Zhang, and Alois Knoll. A neuro-fuzzy solution for fine-motion control based on vision and force sensors. In *IEEE Int. Conf. on Robotics and Automation*, pages 2965–2970, Leuven, Belgium, May 1998.
91. Qi Wang. Extension to the task frame formalism. In *PhD: Programming of compliant robot motion by human demonstration*, pages 107–130. Katholieke Universiteit Leuven, Department of Mechanical Engineering, November 1999.
92. Lee E. Weiss, Arthur C. Sanderson, and P. C. Neuman. Dynamic sensor-based control of robots with visual feedback. *IEEE J. Rob. Automation*, 3(5):404–417, October 1987.
93. S.W. Wijesoma, D.F.H. Wolfe, and Richards R.J. Eye-to-hand coordination for vision-guided robot control applications. *Int. J. Robotics Research*, 12(1):65–78, February 1993.
94. Patrick Wunsch and Gerhard Hirzinger. Real-time visual tracking of 3-d objects with dynamic handling of occlusion. In *IEEE Int. Conf. on Robotics and Automation*, Albuquerque, NM, 1997.
95. Di Xiao et al. Sensor-based hybrid position/force control of a robot manipulator in an uncalibrated environment. *IEEE Transaction on Control Systems Technology*, 8(4):635–645, 2000.

96. Di Xiao, Bijoy K. Gosh, Ning Xi, Tzyh Jong Tarn, and Zhenyu Yu. Real-time planning and control for robot manipulator in unknown workspace. In *IFAC*, pages 423–428, 1999.
97. Di Xiao, Bijoy K. Gosh, Ning Xi, and Tzyh Jong Tarn. Intelligent robotics manipulation with hybrid position/force control in an uncalibrated workspace. In *IEEE Int. Conf. on Robotics and Automation*, pages 1671–1676, Leuven, Belgium, 1998.
98. Tsuneo Yoshikawa. Force control of robot manipulators. In *IEEE Int. Conf. on Robotics and Automation*, pages 1220–1225, San Francisco, CA, 2000.
99. Yu Zhou, Bradley J. Nelson, and Barmeshwar Vikramaditya. Fusing force and vision feedback for micromanipulation. In *IEEE Int. Conf. on Robotics and Automation*, pages 1220–1225, Leuven, Belgium, 1998.

Index

Camera
  calibration 49, 162
  configuration 11
  model 51
  parameters 82
  pin-hole model 48
  relative pose 52
Compliance 22, 32, 97, 105
Configuration
  endpoint closed-loop, ECL 11, 58, 116
  endpoint open-loop, EOL 11, 58, 89, 115, 118
  fixed 59, 118
  variable 60
Contour following
  2D theory 23
  examples 68
  planar continuous curves 89
  simulation 137
  task specification 91
Control
  joint controller 22
  proportional laws 21
  dynamics 30
  feedforward 19, 24, 28, 63, 98, 101, 109
  finite state 108
  force 18, 31, 61, 63
  hybrid 12, 66
  hybrid structure 17, 21
  loop design 30, 31
  position 19
  shared 12, 61, 63, 66, 116
  tracking 19, 63
  traded 12, 65, 115
  vision 19, 30, 61
Curvature computation 45, 144, 150, 155

Denavit-Hartenberg 131
Direction
  feedforward 19
  force 18
  tracking 19
  velocity 18
  vision 19

Edge detection
  ISEF 7, 40, 94
  relative area 37
Experiments
  fixed EOL 118
  planar continuous contour 100
  rounding of corner 110
  shared ECL 116
  traded EOL 115
  visual alignment 82

Force screw 131
Force sensor 33
Forward kinematics 22, 131

High level task description 55, 66, 69, 92

Image Jacobian 10
Image processing 81

contour measurement 94
corner detection 105
implementation 160
relative area 136

Modelling
camera model 51
contour 43, 96, 143
force loop 31
tangent model 44
Tsai's camera model 49
vision loop 30

Resolvability 11

Sensor integration 12
Splines 146

Task frame formalism 12, 17
Tracking error 19, 92
identification 35
Transformation
homogeneous 52, 129
screw 130
srcew 22
Twist 21
Twist space 12

Visual servoing 19
algorithms 81
image-based 10, 73
position-based 10

Wrench 33
Wrench space 12

Symbols

Variables:

C_x, C_y	:	pixel coordinates of center of image	(pix)
F	:	3×1 force vector	(N)
F_i	:	force in i-direction, i=x, y, z	(N)
${}^bH$	:	homogeneous frame representation of frame b	
${}^b_aH$	:	homogeneous transformation from frame b in a	
I	:	intensity (function) or grey-level value	()
J	:	Jacobian matrix	
K	:	control gain	(1/s)
M	:	(3×1) moment or torque (vector)	(N mm)
P	:	3×1 position vector	(mm)
R	:	(3×3) rotation matrix	
R_i	:	rotation matrix for a rotation around direction i	
S	:	(6×6) screw transformation matrix	
SXS_{ra}	:	shifted x-signed relative area parameter	(pix)
T_{ra}	:	total relative area parameter	(pix)
T^{vs}, T^f	:	control cycle period for vision and force	(s)
T_i	:	translational component	(mm)
V	:	6×1 velocity vector or twist	(mm/s - rad/s)
W	:	6×1 force vector or wrench	(N - N mm)
XS_{ra}	:	x-signed relative area parameter	(pix)
XYS_{ra}	:	xy-signed relative area parameter	(pix)
YS_{ra}	:	y-signed relative area parameter	(pix)
f	:	focal length	(mm)
k	:	discrete time step	()
${}^bk$	:	stiffness of object b	(N/mm - N mm/rad)
${}^bk^{inv}$	:	inverse stiffness or compliance of b	(mm/N - rad/N mm)
q	:	joint angle	(rad)
r	:	radius	(mm)

s	:	arc length	(mm)
s	:	continuous domain Laplace variable	(1/s)
s_x	:	scale factor for resampling of scanline	()
t	:	time	(s)
u, v	:	coordinates relative to left-top of image	(pix)
v	:	(3×1) velocity (vector)	(mm/s)
x, y, z	:	distances (along respective directions)	(mm)
x_p, y_p	:	distances expressed in pixels or pixel coordinates relative to center of image	(pix)
x_s, y_s	:	image sizes: width and height	(pix)
z	:	Z-transform variable	()
α, β	:	angles	(rad)
$\Delta\theta_i$	:	orientation tracking error about axis i	(rad)
Δi	:	tracking error in direction i	(mm)
θ	:	angle	(rad)
κ	:	curvature	(1/mm)
κ_1	:	radial lens distortion coefficient	(1/mm^2)
μ_p	:	pixel dimension	(mm/pix)
τ	:	time constant	(s)
ω	:	(3×1) angular velocity (vector)	(rad/s)

Frame and object indices (as preceding subscripts or superscripts):

C	:	contour
COC	:	corrected offset contour
OC	:	offset contour
abs	:	absolute frame or world frame
ap	:	arbitrary point
cam	:	camera frame
cp	:	calibration point
ee	:	end effector frame
fs	:	force sensor frame
obj	:	object frame
t	:	task frame or tool frame

Signal indices (as superscripts):

a	:	actual signal
c	:	commanded signal
d	:	desired signal
m	:	measured signal
ff	:	feedforward signal
ss	:	steady state signal

Control related indices (as superscripts):

f	:	force direction parameter

tr	:	tracking direction parameter
v	:	velocity direction parameter
vs	:	vision direction parameter

Direction or type related indices (as subscripts):

i	:	substitute for either x, y or z
p	:	parameter expressed in pixels
s	:	size
ra	:	relative area parameter
x	:	component along or about x-direction
y	:	component along or about y-direction
z	:	component along or about z-direction

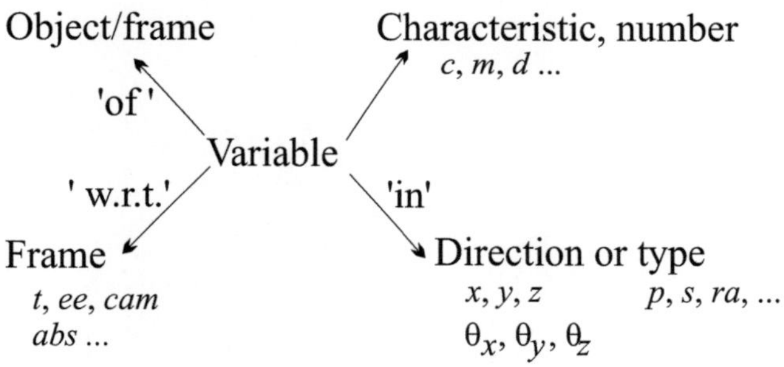

Fig. S.1. Convention for sub- and superscripts

Abbreviations:

CSD	:	control space dimension
DH	:	Denavit-Hartenberg
DOF	:	degree of freedom
DSP	:	digital signal processor
ECL	:	endpoint closed-loop
EOL	:	endpoint open-loop
FWK	:	forward kinematics
IO	:	input/output
$ISEF$	:	infinite symmetric exponential filter
TCP	:	tool center point
TF	:	task frame
VSP	:	video signal processor

Springer Tracts in Advanced Robotics

Edited by B. Siciliano, O. Khatib, and F. Groen
Published Titles:

Vol. 1: Caccavale, F.; Villani, L. (Eds.)
Fault Diagnosis and Fault Tolerance for Mechatronic Systems: Recent Advances
191 pp. 2002 [3-540-44159-X]

Vol. 2: Antonelli, G.
Underwater Robots: Motion and Force Control of Vehicle-Manipulator Systems
209 pp. 2003 [3-540-00054-2]

Vol. 3: Natale, C.
Interaction Control of Robot Manipulators: Six-degrees-of-freedom Tasks
120 pp. 2003 [3-540-00159-X]

Vol. 4: Bicchi, A.; Christensen, H.I.; Prattichizzo, D. (Eds.)
Control Problems in Robotics
296 pp. 2003 [3-540-00251-0]

Vol. 5: Siciliano, B.; Dario, P. (Eds.)
Experimental Robotics VIII
685 pp. 2003 [3-540-00305-3]

Vol. 6: Jarvis, R.A.; Zelinsky, A. (Eds.)
Robotics Research – The Tenth International Symposium
580 pp. 2003 [3-540-00550-1]

Vol. 7: Boissonnat, J.-D.; Burdick, J.; Goldberg, K.; Hutchinson, S. (Eds.)
Algorithmic Foundations of Robotics V
577 pp. 2004 [3-540-40476-7]

MIX
Papier aus verantwortungsvollen Quellen
Paper from responsible sources
FSC® C105338
FSC
www.fsc.org

If you have any concerns about our products,
you can contact us on
ProductSafety@springernature.com

In case Publisher is established outside the EU,
the EU authorized representative is:
Springer Nature Customer Service Center GmbH
Europaplatz 3, 69115 Heidelberg, Germany

Printed by Libri Plureos GmbH
in Hamburg, Germany